Finish Carpentry Illustrated

Elizabeth and Robert Williams

TAB BOOKS

Blue Ridge Summit, PA

FIRST EDITION
FIRST PRINTING

© 1990 by **TAB BOOKS**
TAB BOOKS is a division of McGraw-Hill, Inc.

Library of Congress Cataloging-in-Publication Data

Williams, Elizabeth, 1942 –
    Finish carpentry illustrated / by Elizabeth and Robert Williams.
      p.   cm.
    Includes index.
    ISBN 0-8306-9434-X   ISBN 0-8306-3434-7 (pbk.)
    1. Finish carpentry—Amateurs' manuals.   I. Williams, Robert
Leonard, 1932 –   . II. Title.
TH5640.W55   1990
694′.6—dc20
                                                90-40557
                                              CIP

TAB BOOKS offers software for sale. For information and a catalog,
please contact TAB Software Department, Blue Ridge Summit, PA 17294-0850.

Questions regarding the content of this book should be addressed to:

Reader Inquiry Branch
TAB BOOKS
Blue Ridge Summit, PA 17294-0214

Acquisitions Editor: Kimberly Tabor
Book Editor: Cherie R. Blazer
Production: Katherine G. Brown
Book Design: Jaclyn J. Boone
Cover Photograph: Brent Blair, Harrisburg, PA

Contents

Introduction

*A*s the title implies, finish carpentry deals with all of the tasks that are normally associated with completing the "turnkey" aspects of carpentry work. Rough carpentry by strict definition includes only that work that actually deals with wood cutting, shaping, and fastening. Modern carpentry, however, has been expanded to include all of the work on the house, from the grading for the basement to the installation of the studding, subflooring, joists, rafters, and roofing.

In the rough carpentry phases of your house you had the basement excavated, footings poured, basement or foundation walls erected, and all joists and headers installed for the flooring before the subflooring was installed. Then you had to erect wall frames that included sole plates, studs, top plates and top caps, cripple studs, rough window and door openings, headers for windows and doors, the rafters, ceiling joists, ridge beam, girder or girders, and sheathing for the roof and external walls or wall coverings.

When you reach the finish carpentry stages, you are ready to nail in the finish flooring, lay tiles, install and trim windows and doors, install the finished wall covering, nail in all the molding, install the light fixtures—in brief, all of the jobs, large and small, that are necessary for the house to have a finished and professional look.

The rough carpentry stages of house building require special equipment and skills. The work is generally physically taxing and requires considerable strength and endurance. Finish carpentry work requires a few simple and inexpensive tools and the typical job does not require a great amount of strength. Finish carpentry work does, however, require considerable patience and an ability to take great pains with the work. Any flaws that are created in rough carpentry can often be concealed or covered with paint, paneling, or flooring. Flaws in finish carpentry cannot be covered; they can only be corrected, which often requires a substantial investment of money, physical effort, and time.

In a sense, rough carpentry is similar to cutting a stone, while finish carpentry is like polishing the cut stone and putting all of the professional touches on it. Each of the major carpentry phases will have its demands and requirements, and each has its rewards when the job is done properly.

Rough carpentry requires basic abilities: to lay cement blocks or bricks in a relatively straight line and with a pleasing appearance, to saw along a marked line, and to drive a nail acceptably. Finish carpentry requires a constant attention to the job and the willingness to start over when the work is flawed. You must face your work regularly as long as you live in the house. When you walk into a house that has been finished by a master carpenter, the quality of the work is immediately apparent. When you enter a room completed by a careless workman, the lack of quality is also equally apparent.

Despite the level of skills demanded in finish carpentry work, there is no reason that anyone who is moderately skillful with tools and hands-on work cannot do exceptional finish carpentry. What is needed is advice and instruction in order to approach the job in the proper fashion. Such instruction can be found in this book, which is geared toward the endeavors of the beginner or semi-skilled carpenter.

For years the cost of building has soared. Today, a modest house with three bedrooms and two baths will cost, on average, about $100,000, depending upon location. In some parts of the country, notably metropolitan areas, the cost of a room addition to an existing house can cost more than $30,000, while in other parts of the nation an entire house can be built for the same price. Coastal and resort properties tend to cost considerably more than comparable housing in commonplace neighborhoods, just as the cost of housing in a country club environment might be much higher than a comparable house in another neighborhood.

Costs of building materials do not vary greatly from one part of the country to another, unless the item is a highly specialized one that must be shipped a long distance. In many cases, the major difference in costs of houses lies in the cost of labor.

Throughout the nation a standard rule of thumb is that labor accounts for at least half of the cost of the structure. For example, in the South a mason can be employed for as little as $8 per hour, while similar skilled workmanship in New England might cost twice as much. The same applies to other skilled labor, such as plumbers, electricians, roofers, and painters. What this means to you is that you shouldn't count on saving a great deal of money on materials; your major savings will be in the area of labor, and the more you can do yourself, the more money you will save.

A block mason charges about $1.25 for every block he lays. A skilled mason can lay as many as 500 blocks in one day, which means that you will pay him $500 or more for one day's work. You can learn to lay blocks acceptably in a very short time, and even if your work is not as fast or as expert as the mason's, you can lay 100 or more blocks per day easily. With a little practice, you have saved $100 for your day's work. With persistence, you can learn to lay 250 blocks per day. Your savings will continue to grow as your skills increase. This basic premise will hold true for virtually all of the finish carpentry work you will do on your house. You can save as much as $50,000 on a huge venture, $20,000 on a more modest job.

No matter what the job size, if you can save thousands of dollars, your efforts have been doubly productive. Not only have you saved money, you have also learned a valuable skill that can earn you money for part-time work. You also know the quality of the work. Best of all, you have the great satisfaction of realizing that you have beaten the system, saved your family money, and taken one more major step on the road to self-sufficiency.

Careful planning is one of the major necessities for finish carpentry. There are many tasks that can be interrupted at any point and then resumed later without problems; others should be carried to their completion before pausing. If you are doing masonry work, you must plan your work around the weather as well as your own time limitations and energy. If you must end a day's work while you still have a huge amount of mortar left, you cannot save the mortar and so have wasted the cost of the materials and the labor and time required to mix it.

If you are ''floating'' a concrete section, it is difficult to stop in mid-job without leaving a telltale crack or obvious mark where one day's work ended and another began. Sometimes it is easy to tell where painting was interrupted on a wall expanse, and many painters will tell you that it is far better to complete a wall, door, window, or other surface before stopping work for that day.

Another consideration is your own fatigue factor. Do not try to work beyond your limits, particularly at first. You will find that as you start work you might tire early, but as you continue you will build greater endurance and you will be able to work for hours without rest after a short time.

Still another factor is patience. You might make a particular cut in a length of wood and decide to use the cut even though it creates a bad fit. It is much better to try the cut over and over until you have the perfect fit. The same holds true for sanding drywall. You can apply the compound quickly, then sand it after it has dried a short time. After the spot is painted, however, mistakes are very apparent. Every line, every rough spot in the set compound will be accentuated by the paint, and you will have a roomful of defective jointings or nail holes—and the added problem of removing paint. Spend your time on each and every nail hole and joint and sand slowly and carefully until the spot blends perfectly with the surface. You will then be able to apply a smooth coat of paint that will not reveal the defects.

Do not leave any phase of your work until you are totally satisfied with it. Do not allow yourself to get into such a hurry that you become disenchanted with the work and decide to settle for what is merely satisfactory rather than what is exceptionally good.

Do not lose confidence in yourself. When you see the professional mason or electrician or painter at work, there is a tendency to compare your own abilities, gained during a period of days, to those of professionals who have devoted decades to their crafts. If you see a mason complete several courses in a wall while you are trying to complete part of one course, remember that in time speed and skill will come. Keep working steadily.

A work schedule is an important part of your responsibility. This does not mean that you must set deadlines and adhere to them rigidly, even at the expense of quality work. If you wish to keep your spirits high, however, you will need to see progress. Few disappointments in building are greater than the realization that your work is moving so slowly that you will need years rather than months or weeks to complete your work.

Plan to work a specific number of hours each day if possible, or a certain number of days each week. You might choose to work every afternoon for 3 hours, weather permitting, or opt for devoting every weekend to the job. Whatever your decisions, try to keep a steady work pace going, and try to plan your social life and occupational responsibilities so that you will have plenty of time to work on your house project. One very practical reason for a steady work schedule is that the longer the job requires, the longer your materials will be exposed to weather and vandalism.

When possible, plan your work around the weather. If you know that the winters are too cold for outside work in your area, schedule accordingly. In some areas autumn is very rainy, and you will want to have the house dried in before the rainy season begins. If March is a very windy month, try not to work on framing during this time period and plan to stay off the roof during blustery weather.

Give considerable attention to your help, if you will have any. If you plan to hire part-time help, arrange your work so that you can get a full 60 minutes of work each hour from your paid employees. If there is heavy work to be done, have the work at such a point that your helpers can begin as soon as they arrive.

If you must rent tools and equipment, again schedule your work so that you do not have the rental equipment standing idly while you complete other work. If you must rent a sander, schedule your work so that you can pick up the sander early in the morning and finish working with it early enough in the day so that you can get it back before the close of the business day. Otherwise you might be charged an extra day's rent. Plan other work accordingly.

You can also plan your purchases intelligently. Do not buy any materials before you need them, unless you can get them at a very special price or unless they will not be available later. If your money is tied up in the materials, you run the danger of damage and you lose the interest the money would earn if left in the bank.

Throughout this book all instructions will be provided in very simple, easy-to-understand terminology. No special vocabulary is used unless absolutely necessary or unless it will be helpful to the beginning worker to understand the language of the job. Directions and instructions are geared to the weekend or beginning builder rather than to the professional or experienced carpenter. When possible, directions are provided so that women as well as men can handle the work. Women tend to be very good at finish carpentry.

Directions are also geared to younger as well as to older workers. Many retired or semi-retired workers find great satisfaction as well as a source of extra income by doing part-time work around their neighborhoods. Young men and women, still in high school, have found that they can learn masonry, carpentry, plumbing, and electrical work quite easily. Many excellent roofers are still in their teens or past age 60.

Instructions given here are based on the assumption that the do-it-yourselfers do not have unlimited funds and that they have a desire or need to stay within a budget. When appropriate, general costs of tools are provided so that those unacquainted with equipment costs will be better guided in their purchases. You can buy a carpenter's level for as much as $60 or more, but you can also buy a very serviceable level for as little as $5.99. You can buy a hammer for $6.50, but the hammer is barely useful; for $6 more you can buy an excellent hammer that is well balanced and will last for decades. At times saving money is not economical; genuine economy comes only when you can save money while obtaining excellent quality merchandise.

If you are a total novice and have never before held a hammer in your hand or sawn a board, you can still use this book and let it guide you through the basic steps toward successful finish carpentry. You will gain the experience and knowledge of the authors and other builders and carpenters the authors interviewed while constructing this book.

What is presented here is what the authors have found to be workable. If your own judgment differs, by all means rely on your own wisdom if you feel it will guide you better. If you know a better and simpler way of accomplishing a goal, follow it.

This book includes information on foundation wall finishes, wall coverings for interiors, wall coverings for exteriors, window and door treatments, ceiling installations, roofing, room additions, carports, decks, and tips for better and faster work, with an overall emphasis on how to save money while developing a more self-reliant approach to home construction.

For quick reference, there is a detailed index at the conclusion of the text. There is also a glossary of words and their definitions so that you will not have to thumb through text in order to understand a particular term.

The overall emphasis of this book is on the highly visible work in any house or addition or extension. High-visibility work is any part of the structure that observers first notice as they enter the house or the work area. It is the quality of work that gives the professional, finished touches to the carpentry project, whether it is a window frame or a major structural element of the house.

Instructions are based on the premise that the carpenter or do-it-yourselfer is working alone. We include a variety of suggestions that will enable one person to accomplish nearly every facet of house construction, once he passes the basic foundation or framing work.

Permits are always a part of construction work. If you are not familiar with the permit requirements in your area and if you have not consulted the local building codes recently, do so before beginning your work. Permits are not based on the finest building techniques but on the lowest acceptable levels of materials and workmanship. Anything lower could very easily result in danger to yourself, fellow workers, and family. Building regulations are designed for safety; not to harass you.

Cooperate fully with the inspector when he visits your work site. If he allows you to drop below the requirements, he could lose his job. Do not ask him to take that chance. If you cooperate completely with the inspector, he will be far more likely to cooperate with you and to make your work as easy as possible. It is helpful to call him well in advance so that he can give you a tentative visitation time. This step is particularly necessary if you are building several miles outside the city limits and the inspector will need half an hour or so to make the drive.

When the inspector arrives, make sure you are on the scene and the work he is to inspect is ready. Do not ask or expect him to sit and wait for you to complete the steps. When he makes suggestions or requires modifications and upgradings, take notes. Ask questions. Be certain that when he leaves you will be able to comply with his directives. Be doubly certain that on his subsequent visit you have made all of the required modifications so that this particular stage of your work can be passed.

If you do not meet certain code requirements, you can be compelled by law to take down a certain portion of your house and rebuild using standard materials. Or you can be compelled to take out wall coverings so that electrical connections can be inspected.

When you have passed the basic inspection points and are ready to concern yourself with only finish carpentry, you will have to make decisions that are economically very important to you. When you go to buy windows, you will find that a window for a basement can be bought for as little as $40 or as much as $400. The cost of doors can range up to more than $700. The same is true of shutters, bathtubs, kitchen appliances, shelving, wall coverings, carpets, and light fixtures, as well as numerous other elements of the house.

Note that you can save over $2000 on windows alone if you elect to install less expensive units. You can save $5000 on roofing if you are willing to modify your plans slightly. This is not to suggest that you should buy less expensive materials but only to remind you that you can save alot of money by buying carefully. Do not, however, attempt to save money at the risk of buying and installing materials you will not like later. If you install a basement window, you might find that you cannot remove it later if you change your mind. Be sure that you are buying what is best for you and your bank account.

Now, good luck and good building!

Chapter 1

Foundation wall finishes

*I*f you initially planned to build a house without a basement, you might want to reconsider. From an economic standpoint, a basement and a second floor are among the greatest bargains in construction today. If your main floor costs you $45 per square foot to build, you will find that the basement and second floor can be built for half that cost.

A good basement adds greatly to the economy, convenience, comfort, appearance, and utility of a house. A bad or defective basement is a liability. Consider the advantages and disadvantages of a basement.

If you have a main floor of 1500 square feet, you can double the size of your house very easily by adding a full basement. You can divide the basement into six large rooms and have a much larger and more convenient house. If you plan to heat partially by a woodstove, you can place the stove in the basement and let the heat rise up the stairway. This way, you can have all of the advantages of a wood fire without the mess.

A basement makes a wonderful family room, workshop, game room, storage area, home office space, pantry, laundry area, and space for such extras as a darkroom or hobby room. Because earth provides excellent insulation, a basement is very easy to heat and cool. You can save considerable money, depending upon what part of the country you live in, by using the basement for a family room that can be heated easily in winter and cooled easily in summer.

Cool air from the basement will help cool the main floor of the house, just as the warmth from the basement in winter will help heat the main floor. A basement entry helps protect the remainder of the house from mud and dirt and moisture during bad weather, and you can convert basement spare rooms into guest rooms easily and relatively inexpensively.

If you live in a storm area, a basement is one of the best investments you can possibly make. Recently our own home was struck by a fierce tornado, with winds that peaked at more than 500 miles per hour. The house was totally destroyed, but the basement provided such protection for the family that no one received even a minor injury or scratch.

You can install a canning kitchen in the basement and preserve foods without damaging expensive tiles or other floor coverings in the upper areas of the house. Plumbing pipes can be run through the basement, which decreases the possibility that the pipes will freeze in winter. If repairs are ever needed, they are much easier to make in a basement than in a crawl space under the house.

The major disadvantages of a basement are that leaky basements are a major cause of mildew, mold, bad smells, and the resulting damage to any items placed there.

BEFORE YOU BEGIN

As far as the expenses of adding a basement, you will pay the excavator that you hire to dig the footings in excess of $100 per hour, and this figure often includes travel time. Consider that you will pay $300 or more to have the basement excavated, and for another $200 or so you can have the basement dug.

When you are ready to have the footings poured, you can either mix the concrete yourself and pour it, or you can call a concrete company and have a truckload of concrete sent out. If you choose to call the concrete company—which is the best choice if time is more important than money—you need to have the footings prepared before the truck arrives.

Remember that the footings need to be inspected. Wait until the excavator or grader is about three-fourths of the way through with digging the footings, then call the inspector. He should arrive in time for the final digging. Call the concrete company as soon as you call the inspector so that the truck can start to your house site.

There is a good reason for trying to schedule these visits at the same time. If the inspector finds a problem, he can tell the grader, who in turn can correct it while he is on the site. Otherwise you will have to call the grader back out and pay extra money, unless you have agreed that his fee guarantees footings that will pass the inspector's examination.

Once the inspector has seen and accepted the footings, either you or the grader can start to stake the footings. This involves driving stakes at crucial points throughout the footings to help keep the footings deep enough at all points and level. The stakes can be either measured or marked. The level mark is generally made using a transit, which enables the user to find the exact level mark at all points. If the concrete is poured keeping the top of the concrete level with the tops of the stakes, the footings will be level (FIG. 1-1).

You can save money by laying off the footings marks before the grader arrives. To do so, you must square the house dimensions by using a measuring tape, nylon cord or similar material, and a framing square.

LAYING OFF FOOTINGS

Start by laying off the tentative basement area. Decide first what size basement you want, with the understanding that the basement size will be the house size, if the basement is a full one.

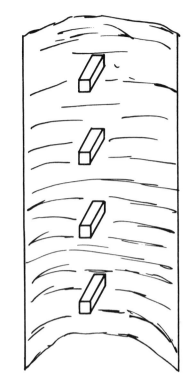

1-1 When the footings are dug and leveled, drive in the short stakes that serve to help level the concrete.

Remember that dimensions should be divisible by four because drywall, paneling, and other building materials usually come in 4-×-8-foot panels. If your house is an odd size, you will need to cut one of the panels lengthwise each time you put up wall covering.

If you want a moderately large house, a length of 52 feet and a width of 32 feet are good measurements. You will then need 13 wall covering panels for the length and 8 panels for the width. You will not have any waste.

Placing stakes

When you lay off the basement, decide approximately where the house is to be located and then put up tentative corner stakes. Place the first stake where you want the front corner of the house to reach. You can use a framing square or any number of other devices to help with the squaring.

Lay the framing square (or a panel of plywood that is cut accurately) so that one corner is against the stake and the two outside edges are aligned with the outside lines of the proposed house. Tie the cord to the stake and pull it 32 feet (or whatever your dimensions are), making sure that the cord is aligned perfectly with the factory edge of the plywood or the edge of the framing square. When you have reached the 32-foot mark, drive another stake into the ground to mark the spot.

Return to the first stake and pull more cord along the other outside edge of the plywood or framing square until you reach the other dimension (in this hypothetical case, 52 feet). Put up the third stake. Finally, put up the cord for the fourth and final side.

Checking for squareness

You must now check to see that the house is squared. First, clear the site of all litter, high grass, or other obstacles. Measure from one corner diagonally across the house expanse to the opposite corner. Be sure the measuring tape is pulled tight and that you measure from the outside corner of each tape (FIG. 1-2).

Write down the measurement and then measure from the other direction diagonally across the house expanse. Compare the two measurements. They should be exactly the same. If they are the same (allowing no more than 1/4 inch margin for error), your basement area is square

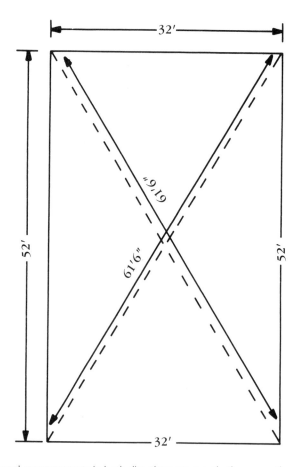

I-2 If diagonal measurements in both directions are exactly the same, the corners will be square.

and you are ready for the grader and his equipment. Do not be disappointed if your first effort is not accurate; in all probability you will need to measure repeatedly before you have a true reading.

There is another way you can measure for squareness. Mark a line or use a straightedge so that you can move along the line until you are 5 feet from the corner. Mark the point. Measure 5 feet down the other side of the corner and mark the 5-foot point. Then measure diagonally across the inside corner from point to point. You should have a 7-foot measurement or very close to it.

Repeat the measurement at every corner. Move stakes in or out until you have four perfectly square corners and the same lengths diagonally across the house site from corner to corner in both directions.

The excavator will dig so that you will have at least 2 or 3 feet excess basement area on each side of the line. Later you will need room to work behind the basement walls in order to seal or waterproof the area (FIG. 1-3).

I-3 In the foreground is the trench for the footings. The grader has left abundant work room behind on all sides of the house so that proper sealing of the exterior walls can be accomplished with a minimum of difficulty.

Marking wall lines

When you have the house staked off exactly, you need to mark the lines for each wall. The grader cannot leave the cord in place but he will need a guideline. For marking, use lime, sand, flour, fertilizer, or any other granular substance that will show up in contrast with the dirt or grass in the basement area. Fill a small bucket with the material and place it near the wall line you wish to mark. Pour a handful in a thick and clearly visible line over the cord, just as if you were marking the baselines on an athletic field.

Mark all four wall lines and then remove the cord. The grader can use the chalk or lime or fertilizer lines as the outside wall line for the house. When the basement is dug, you can use the same process for marking off the footings lines.

The grader will dig the footings on both sides of the chalked line. Your footings should be 18 to 24 inches wide at all points, with many builders preferring the 2-foot widths. If you are going to use 8-inch concrete block foundation walls, the 18-inch footings are usually acceptable. If you intend to use 12-inch blocks, you will want to use 24-inch footings.

Check with local building codes before you specify widths or depths for footings. The figures given here are general, but your state or county might have variant requirements. Footings must be dug until they are deeper than the frost line; otherwise your house foundation could be damaged seriously by hard and deep freezes.

When the footings are being poured, see that the concrete is distributed evenly across the footing area. You can use a shovel to help distribute the concrete. You can also use a short length of 2 × 4 held upright to help you shift concrete until you have it at the top edge of all of the stakes.

BUILDING BASEMENT WALLS

You can dig your own footings if you prefer not to pay someone to bring heavy equipment onto your property. The work is hard and slow, but it can be done.

Digging the footings

You will need only a pick and a shovel for the vast majority of the work. Mark off 1 foot on each side of the center line and dig down 14 inches, or until you are below the frost line or until you have satisfied local building code regulations. Dump the dirt to the outside of the basement area and use it later when you are backfilling the basement.

Start to dig the footings and see what sort of progress you are enjoying. If the work is too grueling or slow, you can stop and call the grader. If not, you can continue until the footings are dug. It is possible to dig footings for an entire house in one weekend and thus save a considerable amount of money, particularly if you do not plan to include a basement.

Ordering materials

When you are ready to build basement walls, you will need to lay off the wall lines after you have ordered your blocks. Most firms that sell cement or concrete blocks will advise you at no charge as to how many blocks you will need. For the basement of a 1500-square-foot house you will need about 1550 blocks. This figure allows for four windows and one door in the wall. If you plan to have no windows or exterior door, you will need another 100 blocks.

Cost of cement or concrete blocks is fairly standard, but you might want to call several companies for comparative purposes. Average cost of a 12-inch block is $1.25, an 8-inch block about $.85.

When you make your initial order, be sure to ask if you can return any unused blocks. If so, order an extra 100 blocks or so. You will need to break some blocks and some will be broken accidentally.

Ask also if the order will include 12-inch halves, grooved blocks, and an adequate supply of end of corner blocks. The usual block is called a stretcher. It has two "ears" on each end and an indented area between the ears. This type of block cannot be used on corners or window or door rough openings. A corner block has one flat end and one "eared" end. A grooved block has a vertical groove that is located precisely where the metal protrusion of a window will fit. You will need a line of these grooved blocks on each side of windows to allow the window framing extension to slip down the groove and prevent the window frame from falling out.

You will need 12-inch halves for nearly all walls. Because 8-inch corner blocks are easily broken, no halves are needed, but it is extremely difficult to break a 12-inch block exactly.

To lay blocks you will need a trowel, a level, a mortar box or pan, two or three mortarboards, a shovel, a 5-gallon bucket, and the ingredients for the basic mortar. A jointer, masonry hammer, and brush are also needed.

You can buy an excellent trowel for $20. A good 4-foot level will cost about $50 or more. You can make your own mortar boards by using a 2-foot square section of plywood and two 2-foot lengths of 2-×-4 stock. A masonry hammer or bricklayer's hammer will cost about $12. If you do not want to buy a masonry hammer, use a carpenter's claw hammer. It won't produce the same results, but it will be adequate.

A cement or concrete mixer, if you choose to buy one, will cost about $500. You might want to investigate the possibility of leasing a mixer rather than paying so much money, unless you plan to use it later for other do-it-yourself jobs.

You can buy a mason's 4-foot level (as opposed to a carpenter's level) for about $35. This level has a mason's scale so that you can use it to determine how many courses of blocks or how many stretcher blocks are needed to fill a certain expanse of wall. You can also buy a complete mason's tool kit for about $45. This kit includes trowel, jointer, chisels, and other equipment.

Mixing the mortar

The mortar you will mix requires three ingredients: water, mortar mix, and sand. Start by putting 5 gallons of water into the mortar box and add eight shovelfuls of sand to the water. Mix well and then add one 70-pound bag of mortar mix. Mix until the cement is fully integrated with the sand and water. Finally, add 16 more shovelfuls of sand to the mixture. If you need to do so, add a little more water. This mixture should give you a mortar that is plastic enough to work easily and thick or stiff enough to support the weight of a block. If your mortar is too weak, the weight of the block will force all of the mortar from the joint, resulting in uneven joints and weak walls.

Laying the block wall

When you begin to lay a block wall, start at the corners. You might want to start at one back corner and, after you have erected the batter boards, place a bed joint of mortar on top of the footings and in line with the line you have stretched.

You can construct batter boards by driving stakes on both outside edges of a corner of the footing and running lines to the next corner where you do the same thing. Stretch the lines in both directions so that they cross and form a perfectly square corner. Do this at all four corners (FIG. 1-4).

When you are ready to lay blocks, the corner block will fit exactly into the angle formed by the two cords or lines. Do not let the block touch the line or you will destroy your perfect squareness. Let the block come as close as possible without actually touching the line.

It is a good idea to form a mortar bed for the blocks, then work at something else for several minutes while the mortar starts to set up or harden. This first course or row of blocks is crucial and you do not want the block to sink into the mortar too far and destroy the level line you have established.

Lay the first block. Then move to the other end of the wall and lay a block for the corner. Position both blocks in the same fashion; that is, let both point in the same direction.

Measure the distance between the inside ends of the blocks and calculate the distance in inches. Then divide the inches by 16, which is the nominal length of a concrete block. Your result will tell you how many blocks you will need for the first course and whether you will have a good fit. You do not want to start by breaking and trying to piece block fragments into the wall. Move the corner blocks in or out slightly until you have room for a set number of whole blocks.

You can now lay the blocks between the corner blocks. Place enough mortar for a good, thick mortar bed. Lay a block by buttering the end and then positioning and leveling it.

See TAB book #3435, *Rough Carpentry Handbook* for detailed instructions on how to lay blocks.

At each new course, cross the end blocks for a good bond. If the bot-

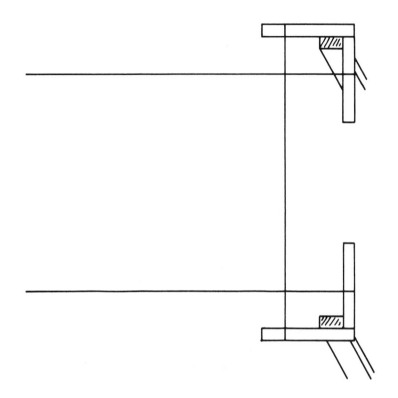

I-4 Batter boards can help set up the perfect starting point for all four walls. Where the lines cross there will be a square corner, and the two outside edges of the corner blocks will fit perfectly into the angle created by the two lines.

tom block is positioned east-to-west, the next corner block should be positioned north-to-south. As you work, keep an eye on the bonding pattern. The vertical joints of alternate courses of blocks should be aligned. If you keep them aligned, you will find that your final block will fit well in each course. If the bond is not kept even, your final block space will be too long or too short. You will then need to try to fit in a section of a block or tighten or thicken the joints of previous blocks. Such efforts require time and energy and duplication of work.

When you come to rough window openings, measure your windows and leave the proper amount of space for metal or vinyl windows. If you plan to use wood windows and frames, allow an extra 2 inches for framing.

A typical door is 36 inches wide, but you will need to allow about 2 inches on each side for the door frame. A standard building practice is to leave a 40-inch-wide rough door opening.

Wall height considerations

The height of your basement walls is a matter of personal preference, but a good arrangement is to lay 12 courses of blocks. You might want to add an extra course for 8 more inches of height. Many people make the mistake of thinking that an 8-foot basement wall will be sufficient, but you must remember that you will pour a cement floor on top of the dirt, which will take 4 inches or more from the height. You will also have pipes from upstairs plumbing dropping into the basement, which will take another foot or more. And you will want to install a ceiling to conceal the pipes and electrical wires, which will take another several inches. The result often is a basement with ceilings so low that you can barely stand erect without bumping your head.

Also consider the stairs into the basement and what the height of the space below will do to the stairway arrangement. If you have architecturally prepared plans, these questions will be answered for you. If you are drawing your own plans, stop and reconsider before you pass the point of no return.

WATERPROOFING BASEMENTS

One of the worst annoyances in a house is a wet basement. The smell is offensive, the damage is extensive, and the use of the basement is minimized severely. The time to stop the leaks is before they start, and as soon as you have finished with the basement wall concrete work, you can start to prevent leaks.

You actually began waterproofing when you left a very thick mortar bed for the first course of blocks in the basement. This thick bed keeps a great deal of water from seeping under the blocks and into the basement.

You added to the waterproofing effect when you poured a concrete floor for the basement. When done properly, a concrete floor will leak or seep very little, if at all. One of the best ways to prevent leaks is to lay a layer of sand, then a layer of gravel, and then cover the gravel with a sheet of thick plastic. Pour the concrete over the plastic and the basement floor is fairly waterproof. Any water that enters the basement at this point must do so through the mortar joints or holes in blocks, or at the point where the concrete floor joins the block wall. One major step you can take to prevent such seepage is called pargeting.

Pargeting

Pargeting is the term for applying parget to a wall. This can vary from whitewash to stucco to a thin coating of mortar or plaster. For basement walls the parget used is mortar, which can be applied so that it closely resembles stucco.

When you apply mortar to the surface of a block, use the inside surface or top surface of the trowel to scoop up a load of mortar. Then turn your wrist, while holding the trowel flat and extended directly over the block when you start. As the trowel descends, turn your wrist to the inside so that the mortar now is pointed at your body. By the time the trowel reaches the block, it is pointed nearly straight down and the blade is vertical. With a quick sweeping motion spread the mortar across the block surface.

When you are pargeting, reverse the entire procedure. Begin by going to the outside of the basement wall and loading the trowel by turning it upside down and cutting the edge into the mortar. Lift the trowel so that the bottom of the blade is pointing upward.

Lower the trowel to the bed of mortar that you applied when you started the wall. As you bring the towel to within a foot of the bed, turn the trowel so that the bottom or back of it now faces the wall, and with a quick jerking motion, sling or dump the mortar onto the set mortar of the first bed. Then push the back of the trowel into the wet mortar and press downward to force mortar into any holes or passageways (FIG. 1-5). Repeat

1-5 Pargeting is one method of waterproofing basement walls.

the process until you have covered a length of wall 5 or 6 feet long. Then start working upward.

Load your trowel, with the bottom of the blade facing upward, and bring the trowel to the wall. Turn it downward so that the bottom of the blade is slightly in contact with the wet mortar. Press inward with the blade at a slight angle away from the wall and sweep the mortar upward. You can press the blade with your free hand to get a smoother and thinner covering.

Keep repeating this process, moving over several inches each time, so that each stroke laps over the previous one by 2 inches. Continue the application until you have covered the entire wall as high as you planned to go. If you are moving slowly, you might want to pause every half hour and dip your empty trowel into a bucket of water and smear the thin coating of mortar or parget with the damp trowel. The result is that you smooth the mortar cover into an attractive stuccolike surface that can then be painted or left as it is.

Pargeting can also help you make use of leftover mortar. There will be times when you are ready to stop for the day but still have a bucketful or more of mortar left. You don't want to waste the mortar because it is expensive and necessary, but you don't want to start a new course or corner. Carry the loaded mortar board to the outside wall and start pargeting where you left off the previous time. You need not worry about starting and stopping. This type of work will not show the marks of the earlier work, and you can apply a square foot of mortar or a cubic yard with equal results. The important point is to dampen the trowel and smooth the surface of the parget before you stop the day's work. If you leave the rough mortar and let it dry, you cannot smooth it later.

NOTE: Whenever you are working with mortar, be sure to cover your work at the end of the day if the weather is threatening at all. Rain can wash all the mortar from joints and you will be left with a severely weakened wall.

Further steps

After the parget has been applied, you will need to coat the parget with waterproofing material. There are many commercial products on the market that you can use. Some of these are black, tarlike substances that can be applied with a very stiff brush or even with a trowel. Other products can be bought in a powder form which, when mixed with water, forms a substance the consistency of thick paint. The "paint" can then be applied with a regular brush.

If you use a brush or trowel for the tarlike products you will find it difficult to clean the tools and they might be ruined. The black waterproofing is unattractive and you will want to conceal it with dirt when you backfill the basement.

The powder substances blend well with the wall and will not prove to be unsightly even if you leave them open to view. Another advantage is that the tools can be cleaned easily.

Allow the waterproofing materials to dry thoroughly. The next step is to install drain tiles. These are long lengths of pipes that are filled with holes on the top and sides so that water seeping into the ground near the basement will seep into the tiles and then flow away down the sloped tiles (FIG. 1-6).

When you have completed pargeting, waterproofing, and installing the drain tiles, you are ready to backfill your basement, unless you plan to install a layer of plastic sheeting over the wall as one final preventative against leaks. At this point you should have a wall that resists all types of leaks.

You can continue applying parget all the way up the basement walls and to the sills if you wish. You will find that the appearance of the pargeted wall is far superior to that of naked blocks. Even painted blocks are noticeably less attractive than pargeted walls.

If you decide to paint the blocks, use a sealer as a first coat. Blocks are permeable, which means that they will allow seepage. You will need to apply several coats of paint to cover the blocks well if you don't use a sealer. The sealer also helps with the waterproofing.

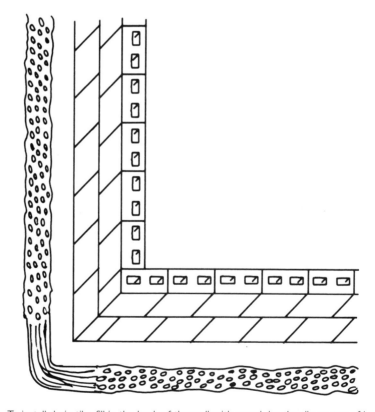

1-6 To install drain tile, fill in the back of the wall with gravel, lay the tile on top of it, then cover the drain tile with several more inches of gravel

COVERING BLOCK WALLS

If you do not want to leave the uncovered blocks visible as part of your basement wall, there are various materials that can be used to face the blocks. Two of the most common of these are brick and stone.

Brick facing

You can buy bricks in so many colors and styles, you are sure to find a style that will harmonize with your house. If you plan to use brick facing for the walls, use 12-inch blocks until you reach the top of the ground. After that use 8-inch blocks that are offset so there will be room for a course of bricks to be laid next to the blocks.

Many building codes require that any blocks used underground must be 12-inch blocks. After you rise above the ground level you can use 8-inch blocks.

Mixing the mortar   To lay the brick facing, mix mortar as you did earlier. Unless you have several hours in which to work, it is best to mix only half a batch of mortar. Bricks do not require nearly as much mortar as blocks, and you will have considerable waste if you have to stop work early. When the mortar is mixed, stack a supply of bricks near your work area so that you can have them handy. This will save time and energy.

Rather than move a mortarboard from one location to another, make new ones. Cut three or four 2-foot sections of plywood squares. Then cut two 2-×-4 lengths 2 feet long for each plywood square. To make a mortar-board, stand the 2-×-4 lengths on a good work surface and place the ply-wood square on top so that the 2-×-4 lengths are about 4 inches from each edge. Drive nails through the plywood and into the short 2 × 4s.

Before you dump mortar on the mortarboards, always wet the boards. The moisture will prevent the mortar from clinging to the mortar-board and starting to set up. Always wet trowels and other tools before beginning to work with mortar. Even the bucket you carry the mortar in should be wet so that the mortar will not stick to the sides of the bucket. If you use a wheelbarrow for hauling mortar, splash water into the barrow before putting in mortar.

Place a small amount of mortar (about 2 gallons, if you are using a large bucket) on each mortarboard after the boards are spaced along the wall where you are working. Then use a block line or its equivalent to establish your brick lines.

Block lines   You can block lines or you can use a makeshift one. The purpose of the block line is to provide you with a guide for the brick positions. If you have to improvise, drive a stake just around the corner of each wall where you are working. Measure to determine the exact height you want the bricks to reach, and fasten the line to the stakes and pull it tight. The line should be slightly higher than the top of the newly laid bricks, so the bricks can barely fit under the line without actually touching it. The principle is exactly the same as that used in block masonry.

If your wall is extremely long, place a brick without mortar on it at the midpoint of the wall so the line will not sag in the middle. Lay the end bricks first so that you establish a good corner height.

Laying bricks To lay the bricks, use the trowel to apply a good thick mortar bed along the entire length of the wall. The bed should be an inch thick if your mortar is relatively thick. When you lay the first brick, set it gently in place so that the end is aligned with the corner of the wall. Push down with slight pressure so that the mortar will be forced up into the holes in the brick. Go to the opposite end of the wall and lay the first brick there so that you have a course height established. Then you can lay the remainder of the bricks in the course.

To lay the remaining bricks, hold a brick in one hand and with the other hand scoop up a small amount of mortar on the end of the trowel blade. Apply a thin "buttering" or mortar on each end of the brick and then place it in position. Press it gently into the mortar bed and butter another brick. Continue in this fashion until the entire course is laid.

When you reach the corner, you will need to start the course that rounds the corner on each end of the wall. Some masons insist—and it is an excellent idea—on setting up corners before they lay any bricks in the main part of the course.

The purpose of the built-up corners is to ensure that a good bond is established from the beginning. The bond creates a tie-in of bricks that will prevent the wall from sagging or even falling. Just as you did with blocks, lay alternate end bricks so that they cross half of the brick in the previous course.

You can buy wall ties that fasten to the block wall. When you reach a tie, press it down so that it is embedded into the mortar bed, then lay bricks over it. The ties will hold the bricks to the masonry wall. These ties are best installed when mortar in the block wall is still wet.

Story pole Many masons like to use a "story pole" for their work. This pole is actually a length of 2 × 4 or other timber that can be used to measure the height of each course of blocks or bricks.

To make a story pole, measure the height of a brick plus a mortar joint, which can be whatever thickness, within reason, that you prefer—usually 1/2 inch or slightly less. Mark a point on the story pole that will indicate the height of each course of bricks from the first to the last.

When you are working on a course, hold the story pole up to the wall from time to time to check on your work. If you see that the joint is too thick, press down on the bricks or take up the last few and remortar the bed joint. Be sure to check the height of each course at the corners. Check all four corners as you work. One of the worst feelings is realizing when two courses of bricks meet one will be extremely out of line. Such work is highly unsightly and mars the appearance of the entire wall.

Continue to set the block line for each new course, and keep building up corners and checking with the story pole. At regular intervals you need to measure the distance between the course you are working on and the top of the wall. You want to make certain that the final course of bricks is even with the top of the masonry wall. If you are planning to install brick facing up the entire wall, the measurement is less crucial, but in all events you should keep an eye on the course heights.

One frequent mistake is to let a brick near the corner touch the line and push it up slightly. Then you will compensate by laying a mortar bed that is too thick and when you reach the middle you will find that both ends are low and the center is high.

A similar mistake is to allow the line to sag, and you will have the illusion that you need to make mortar beds thinner. At the end of the course you will realize that your course of bricks sags badly in the middle. To prevent this from occurring, regularly check the tautness of the line. If it is loose, tighten it and fasten it very securely.

Stone facing

One of the most attractive building materials you can put in a house is stone. It is also one of the most difficult and expensive materials to work with.

You can buy the stone by the cord. It is a better value when bought in volume, but one of the disadvantages is that many parts of the country do not have a bountiful supply of native stone. If dealers must import the stone from other states, the shipping expenses are passed on to the builder. If you live in an area where stone is plentiful, you can often find wonderful building stones around stream beds and in fields after the plowing is done. The advantage of bought stone is that you will have a uniformity of thickness. Fieldstone is not uniform at all and you might have to shape the stones individually before you can use them in a wall as a facing.

If you are facing a block wall, separate your stones into piles of roughly the same size and thickness before you start laying stones. You want stones as flat as you can get them, but they must be thick enough to have a face that can rest on the tops of the blocks underneath. As you work, you will quickly start to make the proper distinctions between thicknesses, and you will wind up with three or four piles of stones.

Before you start to lay stonework, you need to know whether your stones are absorbent or nonabsorbent. Absorbent stones are porous and will soak up the water in the mortar and make it too dry to hold or create a bond. Such stones need to be thoroughly wet before you begin work. Nonabsorbent stones will not absorb water so readily, so you do not need to wet them before you start to lay them.

Methods of laying stone   If you begin with large stones at the bottom and work upward with smaller stones, the wall will slant inward very slightly. This slight slant is fine as long as it slants toward the house. If larger stones are used as you progress upward, the slant will be outward. This is not only unsightly but dangerous; the wall will not be stable.

Although it sounds not only strange but impossible, many stone masons like to start at the top of a wall and work their way down. In order to keep the top stones from falling, they put a shovelful of cement into the mortar mixture. When they place a stone they coat the back side with mortar and cement mixture, place it in position, and—while holding it in place—turn it one-third of a turn. The result is a suction effect that holds the stone in place until the mortar starts to set and the bonding is established.

If you try this method, do not attempt to use large stones at the top. Make sure that the stones are not only small in circumference but thin as well. The quick bonding method cannot hold a really heavy stone until the setting occurs.

The argument in favor of starting at the top is that each time you install a stone, mortar will be pressed out and will fall down the wall. If the bottom stones are already laid, the mortar will stick to them and you will have a large cleanup chore after your work is done.

If you do not use this method and plan to work from the bottom up, lay lower courses around the entire house instead of trying to lay stones for one whole wall. Then the earliest laid stones can be settling and the mortar can be curing. You do not want to lay heavy stones over "green" or uncured mortar unless the stones are seated solidly and there is no danger that they can be dislodged easily.

Tools To help with stone shaping, you will need a mason's hammer and chisels. You can also buy a blade for your circular saw that will cut stones. The work is very noisy, very slow, and very dirty. If you use one of these blades, wear protective eyeglasses at all times. The flying stone fragments can be very dangerous.

Mixing the mortar For ordinary stonework, use a ratio of 2.5−3 parts sand to 1 part portland cement, plus whatever amount of water is required to achieve a good plasticity. If the work you are doing will have to support heavy loads or withstand high winds or similar stress, use a ratio of 1 part portland cement to 3.5−4 parts sand. Many people do not realize that a mixture of pure portland cement and water is not as strong as sand mixed with the portland cement.

Laying stones Unlike brick and blocks, which are uniform in size, no two stones are exactly the same size and shape. When you pick up a stone to place it, lay it with its broadest face down and resting on the stones beneath it. You will find that it is also very difficult to achieve a good bond with stones that you could readily achieve with bricks and blocks.

In brick and block masonry you need to butter each unit by applying a small bead of mortar on the ends of the unit. Stones are shaped in such a way that buttering is too difficult to be practical, so a process called "slushing" is used. This procedure is so named because the mortar used has the consistency of partially melted snow.

When a stone is ready to be placed, use the trowel to sling mortar into the area where the stone is to be located. When the entire location is mortared, set the stone gently into position and press it firmly so that it is embedded in mortar. Remove all excess mortar that is squeezed out around the perimeter of the stone. If there are areas that are lacking mortar, fill these in by lifting a small amount of mortar on the edge of the trowel and packing it around the stone.

You will encounter stones that are shaped in such a way that they cannot fit snugly into the available space. When this happens, you will need to embed smaller stones in the wide spaces left vacant by the unsatisfactory fit of the first stone.

Place the largest stones at the bottom. These must support much of the weight of the later stones. As you work your way upward, gradually reduce the size of the stones so that the smallest stones will be saved for the top. The only exceptions are those very small stones that will be inserted in large gaps left between larger stones.

Another distinction between brick or block masonry and stone masonry is that with stones you do not have well-defined courses. You will need to try to maintain courses, but the size and shape of the stones will not permit exact course work. For similar reasons you will not need a block line or story pole in stone masonry. Make visual examinations of each semidefined course in an effort to keep the stones basically on a level, but precise work is impossible.

When laying blocks and bricks, use a jointer for tighter joints. This is a curved piece of metal that will permit you to rake out mortar that had not yet hardened. Sometimes the process is called ''rubbing'' the joints. The purpose is twofold: to make the joint work more attractive and to push the mortar firmly into the joints so that it is more compact and therefore stronger.

If you do not have a jointer for the common masonry work, improvise by using a brick held so that one edge rubs the joint and rakes the loose mortar out to a depth of about 1/4 inch. You can also use a short piece of 2-x-4 stock for this purpose. You can even use your finger, but wear a thick glove; the blocks and mortar can painfully abrade your finger.

In stone masonry you do not joint. The mortar is pressed out so that you have a rounded bead of mortar surrounding the stones. This bead is often left both for appearance and support.

You can face the entire foundation wall of your house using bricks, blocks, or stones. You can face chimneys in the same manner. Once the foundation walls are completed, you can turn your attention to interior work.

Chapter **2**

Wall coverings for interiors

*I*nterior walls can be covered in several ways. The most common wall coverings are drywall that is painted, paneling, wallpaper, and artificial masonry materials. Even fabrics, carpets, and tapestries can be used.

DRYWALL

Drywall, also known as plasterboard, sheetrock, or gypsum wallboard, is usually made in panels measuring 4 × 8 feet. If you have laid out your overall house dimensions so that they are divisible by four, you will have no trouble installing these panels. The edges of the panels will match the studs in your wall framing and you will have a good nailing surface. You will not have to trim or cut the panels, nor will you have to install or move studs in order to have a backing for nailing.

If you choose to install drywall, purchase special nails at the time you buy the panels. Notice the panels have one perfectly flat side and one side that has small offsets around the edges. The panels should be installed so that the offset side is facing outward into the living area of the room.

Drywall panels are very heavy and somewhat fragile, so installation is easier if you have someone to help you lift and carry and hold them. If you must carry panels alone, gently lower the panel until it is resting on one long edge. Lift one end until you can get one hand under it and the other hand on the top edge. Lift the panel carefully and transport it to its destination. Once there, put it down slowly so that the edges do not strike the floor. Do not let it rest on any boards, tools, or other items that might be on the floor.

Nailing the panels

When you are ready to nail the panel in place, measure first to see that you have a full 8-foot height for your wall. If you see that the fit is going to be tight, do not attempt to wedge the panel into position. Instead saw off an inch or so from the bottom edge of the panel. If you try to force the panel into place, you will cause the top edge to break or crumble, and this can never be repaired to your total satisfaction. Remember that you will have baseboards and/or ceiling molding that will conceal any slight gaps resulting from an imperfect fit.

When you are getting ready to nail a panel, stand it on the bottom edge and gently lift the entire panel. While it is in the air, push the bottom edge into position. Then lean the entire panel forward until it is seated against the wall studs. Be sure that you have a good nailing surface behind the panel.

If you cannot remember where the studs are located, you might want to make a slight mark to represent the center of a stud. Use a pencil or small piece of masking tape placed on the ceiling joists and subflooring so that you can readily locate the hidden studs. This is important because when you drive nails into the drywall you want to be able to hit the stud readily without making unnecessary holes in the panel.

Fitting corners

The proper place to start installing drywall—or paneling or any other wall covering panels—is a corner. If your corners are square, you will have no difficulties. If corners are not square, you need to make necessary corrections at this point.

One easy way to check the squareness of a corner is to hold a framing square in the corner with the heel pushed against the wall and the tongue and blade of the square extending along the wall in a horizontal fashion. If the reading is acceptable, hold the square so that the heel is in the angle formed by the ceiling and wall. If both blade and tongue are flush against their respective surfaces, the reading is good (FIG. 2-1).

You can try the same test by placing the heel on the floor and pushing the heel against the wall. Again, the blade and tongue should be flush against their surfaces.

Determining a problem is only the first step. The second step is to correct the problem. The time to do this is when the first panel is installed.

When you push the panel into the corner, you might find that the inside edge of the bottom is flush against the framing but the inside edge of the top extends away from the corner by as much as 2 or 3 inches or as little as 1/2 inch. Attempt to solve this problem as follows:

Check that the wall frame is plumb by holding a level against the wall. If the bubble is not centered, the room is not plumb. You can also hold the paneling in place and hold the level against the outside edge of the panel. Adjust the panel until the level shows that it is plumb.

With the panel in place, hold it by leaning against it and use a measuring tape to determine the extent of the problem. If the top edge is 2 inches off, lower the panel and lay it face down. (The face is the side that will be seen when the walls have been completely covered, so do not place it on

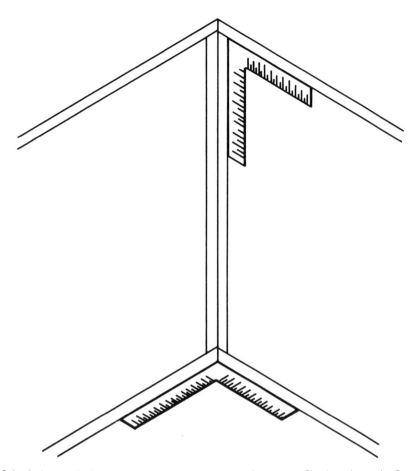

2-1 Before beginning wall covering, the corners must be square. Check angles at the floor and ceiling.

a surface that will mar or damage it.) Remember that the corner was 2 inches off square, and the problem was at the top. This means that if the top is to be moved over until it is flush with the corner, you need to cut off the bottom of the side edge of the panel. Use a ruler or tape or square and mark a point 2 inches from the inside edge of the panel. Make the mark on the bottom edge. Then use a chalkline or straightedge and draw a line from the 2-inch mark at the bottom edge to the corner at the top (FIG. 2-2).

You turned the face down because when you are sawing with a circular saw the blade turns backward—that is, the teeth cut with an upward motion—and the top surface of the wood splinters. If the face is turned up, then the splinters will be on the face surface of the panel and damage it. With the face down, the face side will have a smooth, even cut. However, if you are using a handsaw, mark and cut with the face up. A handsaw does the cutting on the downward rather than upward stroke and the splintered edge will be the one facing away from the carpenter.

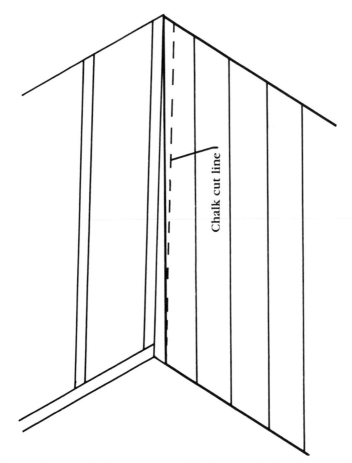

Chalk cut line

2-2 If the angle is wrong in the corners, stand the first panel in place to determine the extent of the discrepancy. Then mark the panel and cut along the chalk line.

Cutting the panels   When you are cutting drywall, you will find that a sharp pocket knife works as well as, if not better than, a saw. To make the cut, lay the panel on a firm, solid surface.

After the cut line is marked, lay a straight edge (a length of true 2-x-4 or 1-x-4 or 1-x-5 stock works well) along the cut line. Start the knife blade into the drywall panel at the top. Press down until the blade has penetrated fully and then pull the knife toward you with pressure maintained until you reach the end of the cut line. When you reach the end of the panel, return to the top and again draw the knife blade along the cut line. Repeat this until the panel has been fully cut.

When the elongated triangle is cut from the edge, the panel should fit neatly into the corner. Stand it in position again, and while holding it with your body, sink several nails through the panel and into the studs. Three nails on each side should hold the panel well so that you can complete nailing.

Use one nail every 2 feet along the top and side edges of the panel. Then move to the hidden stud area and repeat, again using one nail every 2 feet. When the nail is nearly flush, soften your hammer stroke so that you do not damage the surface of the panel excessively. Be sure to sink the nail fully until the head is slightly below the surface of the panel face.

Continue around the room by measuring and cutting the panels to the necessary height and installing them. You should not have to make any more modifications on the sides of panels along the entire wall.

Working around windows and doors

The most troublesome spots in a typical room are the windows and doors, as well as other wall interruptions such as thermostats and wall switches or receptacles. If you do not get a good fit in these areas, you will have a wall surface that is unsightly and that is not energy efficient.

There are two basic methods of fitting drywall or paneling around doors and windows. Method 1 involves measuring, marking, and cutting out for the wall interruptions. Method 2 involves installing the panels on both sides of the windows and doors, then cutting sections to fit over and under the openings as needed.

Method I

When you measure from the last installed panel to the near edge of the window framing, assume that the panel will overlap the edge of the window framing by 11 inches. All you need to do is measure from the ceiling down to the top of the window frame and from the floor up to the bottom of the window frame. If you have 12 inches at both points, lay a drywall panel face up, then measure down from the top 12 inches and mark the point. Do the same with the bottom measurement. Next, measure in from the edge by 11 inches at both of the marks you have just made.

Mark the cut lines as you did before. Use the pocket knife or similar cutting instrument and cut out the rectangle from the panel. When you stand the panel in position, it should fit snugly against the window, with 11 inches extending past the edge of the window.

If there is not a stud or cripple stud to which to nail the inside edges of the panel, you might need to cut off enough of the panel to make it fit well over a stud. Do not install panels that do not have studs behind the edges. If you do so, whenever there is excessive pressure against the panel there is a possibility that it will break if there is no support behind it.

Method 2

To install panels at a window, measure the distance between the last panel's edge and the edge of the window. Then mark the panel and cut accordingly so that the partial panel will fit in the space. Go to the other side of the window and install full-size panels to cover the rest of the wall. When you are ready, measure the distance from ceiling to top of window and between panels. Cut a section of panel to fit the space and install it. Do the same below the window. You can also use this method for hanging drywall above doors.

Try to have the edge of one panel or section of panel snugly abut the edge of the ones on each side of it. You will be able to use compound to fill spaces later, but it is better if the drywall fits well to begin with.

NOTE: You might wish to use one panel of drywall for several of your partial needs, rather than cut from several panels. When you cut off the first segment, set the partial panel aside for later use.

Working around other wall interruptions

While many walls in a home are flat and straight, others have such interruptions as wall switches, receptacles, and thermostats that you need to incorporate when you start to cover the walls. Your problem is that you want the best possible fit but you also want to have the switch box or receptacle ready for use.

A simple solution to incorporate these interruptions is to use chalk. At most hardware stores you can buy a small container of marking chalk that is used with chalk lines. When you are ready to work, empty a small amount of chalk onto a piece of cardboard or similar surface. Dip your finger into the chalk and outline the switch box. Be sure that you have a good covering of chalk.

Stand your drywall panel against the switch box or receptacle (with covers removed). When the panel is in place, push in gently in the area of the box to be marked. The back of the panel will be in contact with the chalk, and when you take the panel down there will be an outline of the switch box or receptacle.

Lay the panel face down on a clean surface and use a knife to cut out the rectangle. When you put the panel back in place the fit around the receptacle should be snug and neat.

If there are no stud backgrounds near the receptacles, you might want to install short or cripple studs or bridging between studs so that the drywall will have a good background support. If for some reason your cut did not work out, set the piece aside and use another. The damaged panel can be used for areas above and below windows and above doors (FIG. 2-3).

Sanding and filling drywall

One of the most important and patience-testing tasks in the entire house is that of sanding and filling drywall. Where you have driven nails, there will be a small indentation. Where every two panels are joined, there will be offsets that are not as thick as the remainder of the panel. You will need to use spackling compound to fill these low spaces and bring them back up to the original surface of the panel. This plaster paste is available as a powder for mixing with water, or premixed.

When you start to work you will need a container of compound, a putty knife, some joint tape, scissors, and coarse and fine sandpaper. Start

2-3 Use partial panel above doors and over and under windows. On the ceiling, note the compound covering the drywall nails and the seams between panels.

by using the putty knife to apply thin layers of compound to the nail indentations. Press the putty knife down so that the compound is packed inside the indentation. When it is filled, smooth the compound as well as you can. Do this for every nail head in the entire panel and in the entire room.

When you have finished, allow the compound to set until it is firm and hard. Do not attempt to work it in any way while it is in a plastic or semiplastic state.

Filling the joints is accomplished in an entirely different manner. Start by using the putty knife to spread a thin layer of compound down the joint area. Do not fill the offset. When this is done, measure and cut a length of tape that will reach the entire length of the joint. You might wish to start at a short joint to test your skills.

After you have cut the tape, start at the top of the joint and press the end of the tape into the compound by using the broad edge of your putty knife. Load the putty knife with compound and smear the compound over the tape, pressing down so that the tape is pressed into the compound bed and so the new compound is pressed into the tape. Finally, use the putty knife to taper the compound until it is smooth at the edges. When you are finished, the level of the joint should be equal to that of the rest of the panel (FIG. 2-4).

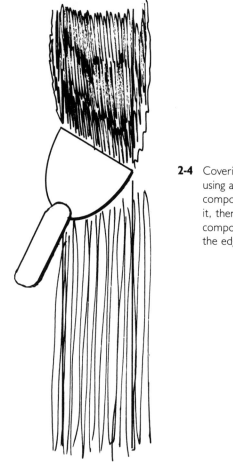

2-4   Covering seams between drywall using a putty knife. Apply compound and embed the tape in it, then taper or feather the compound until it is smooth at the edges.

In both nail heads and seams filling, *feather* the compound for neater and easier work. This means that as you spread the compound, work from the center to the outside, and as you do so you gradually thin the compound until finally it is only a smear.

Do not overload the joint with compound. When it dries and sets you will need to sand it, and the more you pile into the seam the more you will need to sand. Make sure that your seam is filled completely so that you will not have to add more later.

After all joints and nail holes are filled and the compound has set overnight, you are ready to sand. Use a mask to cover your mouth and nose so that you will not breathe the particles of compound; they can be damaging to nasal passages and lungs.

Start with the coarse sandpaper and rub in a circular fashion when you are sanding nail indentations. Start with a very small circle and gradually expand it until you have covered the entire expanse of compound. If the sandpaper is abrasive to your fingers, wrap the sandpaper around a small block of wood. A 6-inch piece of 2 × 4 works well.

When you are sanding the joints, start in the middle and pull the sandpaper toward the outside, or use a side-to-side motion. The air will soon be filled with motes of compound dust, so you will need to ventilate the area if it is enclosed.

After you have used the coarse sandpaper or emery paper to reduce the compound level to that of the rest of the panel, use fine sandpaper to eliminate scratch marks and any irregularities that still exist. When the compound is smooth, run your fingers over it to see if you can tell where the compound stops and the panel surface begins. If you can tell, you need to sand more.

After the entire room has been sanded, use a vacuum cleaner to collect all of the dust. Use a damp cloth to wipe the surface of the panels, and a stiff-bristled brush to clean away all of the particles of compound.

PANELING

One of the most popular types of wall covering is paneling, which is usually 4 feet wide and 8 feet high. Like drywall, you can buy longer panels, but the typical room is only 8 feet high and longer panels are not needed.

The popularity of paneling is due in part to the fact that it is easy to work with, lightweight, attractive, and inexpensive. It is possible to find paneling on sale as cheap as $6 per panel, although prices range to $20 per panel or higher. A medium price is $14.

NOTE: Many times when paneling is on sale for very low prices, the materials might be a close-out product that will not be available in the future. If you miscalculate your needs or damage a panel, you might not be able to find more like it. If you overcompensate and buy more than you need, you probably will not be able to return it for a refund.

Ask the dealer if you will be able to buy the same paneling several weeks later, and if he will accept any panels back as long as they are not marred or damaged. Ask also what percent of refund his store gives. Some dealers have a policy that they will refund only 80 percent of the cost of an item. The explanation behind this policy is that often there are blemishes to the materials that are not discernible at first sight but which will appear to later customers who will refuse to accept the merchandise. Other factors include the cost of bookwork before the refund can be made. Dealers reason that it costs them as much in manpower to refund money as it does to take in money on sales. If you choose to return merchandise, you will be expected to provide the truck and manpower to load and haul the materials.

Once you have made your paneling selection and are ready to work, follow the same preliminary steps as you did when installing drywall. Check for squareness and plumb and make the necessary adjustments to get a good corner fit.

Paneling installed over drywall gives a sturdy wall that provides good insulation and soundproofing. Paneling is usually very thin and offers very little in insulation value. If there is no support behind the paneling, it tends to give under pressure.

Using both drywall and paneling doubles the cost of the wall covering, but at the same time can be economical. If you measure the savings in cooling and heating over the years, the added thickness of the wall can pay for itself. Although paneling does not have great insulation value, it adds a great deal to the appearance of a room.

Installation

Start paneling in corners the way you did with the drywall. Make certain the first sheet of paneling is modified to fit the corner and will still leave you with an outside edge that is perfectly level or perpendicular. You can determine the vertical trueness by holding a level against the outside edge while you hold the paneling in place in the corner.

When you are nailing up paneling, tap the nails lightly. You will not be able to conceal any scars or blemishes you inflict upon the face of the paneling. A good idea is to stop tapping just before the small head of the nail makes firm contact with the wood. Do not sink the nail firmly as you normally would.

You can buy paneling nails that are especially made for this type of work. The nails are thin-shanked, short, and grooved. These grooves will keep the nails from pulling out later, so you do not need to sink them so vigorously. You can also buy nails that are tinted to match the primary or predominant color of your paneling. These tints range from almost pure white to dark brown. When you buy the paneling, buy the nails at the same time to be sure that matching tints are available for the paneling you chose.

When you saw paneling, remember to place the sheet face down so that splintering will be visible on the back side only. Any time you need to lay paneling on its face, make sure you don't lay it on nails, tools, or other items that will mark the finished surface of the face.

Install paneling around light switches and receptacle boxes the way you did drywall. Use the same principles of cutting from the same panel when you need small lengths of paneling.

Cutting the panels

You have noticed that paneling is manufactured with grooves or seams that simulate board edges. These seams are often only 1/8 inch wide. If you ever need to cut a narrow section of paneling, you can cut it along the seam easily. Note also that the widths vary on each unit of paneling. You can usually find the width you need by measuring several of the board simulations singly or in combination. Remember that if you abut two pieces of paneling without a seam between them, the result is unsightly. Incorporate the seam or groove, or at least a part of it.

Incorporate the seam in one of these two ways: Cut the paneling so that your knife or saw is as far to the left or right as you can get it, so that you have a full width of seam on the piece you are using. Or, cut down the center of the groove so that both pieces will have a small groove.

If you choose the latter and rip down the middle of the seam, your greatest problem will be achieving a straight cut. If you do not maintain a straight line, your deviations from a true cut will be very apparent, particularly when a second piece of paneling with a ripped groove abuts it. You can see the mistakes all the way across the room.

You might have greater success if you make the cut so that the entire seam is retained. It is much easier to make a true cut in this fashion. To do so, start the cut with your saw or knife at the very edge of the seam. As you cut, keep the blade against the simulated wood all the way. The result will be a seam that can be joined with another piece of paneling and the juncture will barely be noticeable. The only problem here is that you now have no seam at all on the piece that was left over, and if you plan to use it later, you will have to do so with a second piece that was cut so that the seam was retained.

When ripping paneling in this fashion, the face will have to be up, and it will not be feasible for you to use a circular saw or any other type of power equipment. You can use a handsaw or a very sharp knife with a strong blade. The seam materials cut very easily.

When cutting, place the panel so that the majority of it is supported by the work surface. Allow the panel to extend over the edge so that the seam you plan to cut is barely off the surface. If you need to hold the panel steady and have no one to help you, use a C clamp to hold the panel. Use thickly folded newspaper or similar buffer substance between the clamp and the face of the panel to prevent the clamp from marring the face of your paneling.

MASONRY WALL COVERINGS

Among the reasonably new wall covering products on the market today are the masonry look-alikes. While these products have been available for years, improvements have been made to make them look very much like brick or stone work.

The artificial bricks are made of gypsum and fiberglass. Gypsum is the same material that is used in drywall. You can also buy stone look-alike wall coverings made from gypsum and acrylic. Gypsum is also used to make plaster of paris and other plaster materials.

The look-alike stones are shaped irregularly, just as genuine stones, and they come in a variety of sizes and thicknesses. Although they look like stones that have been cut for construction, they cost a great deal less. You can cover 1 square foot for less than $2.

Look-alike stones are applied with an adhesive that takes the place of mortar. One gallon of adhesive costs about $13, and you will need about 10 gallons to do a wall that is 20 feet by 8 feet.

The bricks themselves are usually slightly more than 8 inches long, about 2$\frac{1}{2}$ inches high, and $\frac{3}{8}$ inch thick. To install, you must first prepare the wall by wiping it clean of all grease or similar soilings, then eliminate all obvious moisture from the wall surface. For best results the wall should be as level as possible.

Installation

Start by using a trowel to apply adhesive to the wall. Apply it in much the same way that you pargeted a wall with mortar, as described in Chapter 1. Cover the wall fully. Do not leave any spaces without adhesive covering. The bricks will stick only where there is adhesive, and if you spread the adhesive too thin the bricks can pull loose. A good covering is about the thickness you would use for extra-thick paint.

When the wall surface is covered, use a putty knife to spread the adhesive on the back of the bricks. Cover each brick completely. When you set the brick onto the wall, push it into the adhesive and work it back and forth slightly to achieve a good bond.

After the bricks are laid, grout the mortar joints with a narrow brush. Apply one coat of sealer if the wall is in a room where there is a danger of grease or water hitting the surface. You can buy the sealer for about $20 per gallon. It is also available in quart cans.

Apply the sealer by spraying it (use an insect spray applicator or spray bottle), or use a paint brush. When the sealer dries it will have a clear finish and will not alter the color of the bricks.

These brick look-alikes can be used outdoors as well as indoors. You can even use them to face concrete block walls or chimneys, or to face indoor fireplaces. Do not use these or any other type of wall covering product inside a fireplace. The bricks are not heat resistant. Use only fire bricks inside fireplaces or woodstoves.

If you use the bricks outdoors, apply two coats of sealer. Apply the first coat, then allow several hours for drying before you apply the second coat.

You can buy a less expensive look-alike brick but it is only suitable indoors. These bricks cost as little as $.80 per square foot of wall space.

The appearance is usually not quite as realistic as the more expensive brick, but the effect is still pleasing and quite satisfactory.

When you are installing these bricks, maintain a course line just as you would if you were laying real bricks. The major difference is that you can start at the floor and use the floor line to get you off to a level start. You can also measure from time to time on both sides of a course to assure yourself that you are still holding the desired course level.

FIBER WALL COVERINGS

One of the most popular fiber wall coverings on the modern market is made from a plant called sisal hemp. Sisal is a white fiber made from the hemp plant. The fibers themselves are very strong—so strong that they are also used for making cords and ropes—and they are very durable. They will last for years when used as wall coverings.

These sisal hemp wall coverings are made by weaving fibers into a tight pattern. The patterns come in a variety of colors and are easily cleaned and also easily installed. Unlike wallpaper and paneling, sisal hemp fibers are tough enough to resist scratching and abrasions. Such wall coverings are ideally suited for any room where there is a great deal of activity or large gatherings.

Calculating needs

The fiber products are usually sold in panels or small rolls that are 2 feet wide and 8 feet long. One roll or panel sells for about $16, and it will take 24 rolls to cover a room that is 12 feet × 12 feet—slightly less when accounting for door and window openings.

Before making a purchase, calculate the number of square feet in a room and divide that number by 16, which is the number of square feet in one roll of fiber wall covering. Then calculate the number of square feet taken up by windows and doors and subtract that amount from the original square footage.

Here is an example: If the room is 18 × 20 feet, you will have two walls that are 20 feet long and 8 feet high, which means you have 320 feet in the two larger walls. The two 18-foot walls have 144 square feet each, which gives a total of 288 square feet. The entire room has 608 square feet.

The room has two windows, each 4 feet wide and 5 feet high. The two windows account for 20 square feet each, or 40 square feet. The door adds 23 square feet of wall space. The total square footage of windows and door is 63 square feet. Deduct the 63 square feet from the previous total of 608 and you have 545 square feet of wall space to cover. Divide the total by 16 (the number of square feet in a roll) and you learn that you need slightly more than 34 rolls of wall covering.

You probably cannot buy a partial roll, so you will need 35 rolls. This will cost a total of $560 if the rolls cost $16 each. You will also need 17 quarts of adhesive to cover the entire room. The adhesive can be bought for $4 per quart, and the total cost will be $68. It will cost about $628 to cover all of the walls. This is comparable to the cost of paneling and sisal will give you a distinctive wall covering that will remain attractive for many years.

Installation

Fiber wall coverings can be installed over concrete blocks, wood, paneling, or many other surfaces. First clean the wall of all grease and similar substances. Grease prevents the adhesive from bonding to the wall, as does dirt and other substances. Moisture also prevents bonding, so the wall surface should be dry when installation begins.

When the wall is properly prepared, shake the adhesive can thoroughly and open it. Paint on the adhesive as you would thick paint or paste. Start in a corner and spread the adhesive well. Cover a 2-foot section fully and expand the coverage at least 2 or 3 inches wider than the actual strip.

Unroll the wall covering and, beginning at the top of the wall, position the end of the roll carefully upon the adhesive and press down firmly. The top edge of the roll should align with the ceiling exactly. You can use a level held against the side of the roll if you want to make sure your work is properly vertical. Slowly unroll the panel as you work your way downward. Press the fabric into the adhesive firmly so that a proper bond occurs at every point.

Move 2 feet farther over and apply adhesive again, repeating the process until you have completely covered the wall. You will need to work your way around light switches and receptacle boxes as you did with drywall and paneling. If you have excess fabric when you reach the end of a wall, trim it with a very sharp knife.

You can install fabric easily around window and door frames. When you find that two rolls lap slightly or that one roll laps over the window or door trim, cut it in place. Place the knife blade firmly against the edge of the window frame and cut with firm pressure. Never let the blade move away from the frame. When the cut is complete, press the fabric into the adhesive and that part of the job is done.

Save one roll just for cutting pieces to fit around windows and doors and in other unusual places. Installing fiber wall covering moves rapidly and the work is generally clean and enjoyable.

PLANK PANELING

Many people prefer real wood rather than the simulated wood in 4-×-8 foot sheets of paneling. You can buy real wood planks that are a combination of 6-inch widths and 8-inch widths and 8-foot lengths. These planks are made of pine and each plank or board is machined for a sure fit. The boards are only about $3/8$ inch thick and cost about $1.25 per square foot. Look for these planks at your local lumberyard or building center, where you have a large selection of lengths.

Installation

You can install the planks over existing walls by applying the adhesive as described previously, then pressing the planks into place. As always, start in a corner and be sure that you have a vertical reading when holding a level against the outside edge of a plank pushed firmly into the corner. If you need to trim the corner plank in order to achieve a perfectly vertical reading, do so, or all of the other planks will also be slightly out of alignment.

Spread the adhesive, place each plank in place, and press firmly on the plank until it is seated well in the adhesive. Hold the plank in place for several seconds so that bonding can begin.

Furring strips

If you already have an existing wall and want to nail the planks in place, you will need to install furring strips. These are thin lengths of wood that reach from ceiling to floor and have horizontal strips connecting them. Before applying the strips, run a chalk line down the wall, making sure the line is vertical. Then glue or nail the strips, depending upon the background surface. If gluing, use a glue gun, squeeze tube, or paddle stick to apply the glue along the line (FIG. 2-5). Allow it to set several minutes.

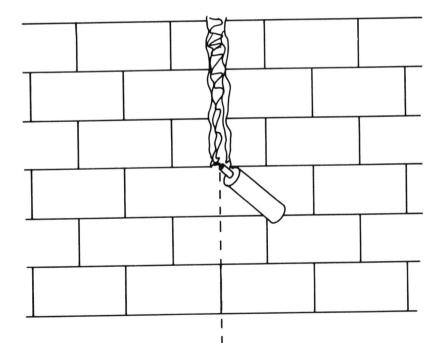

2-5 Before applying lath or furring strips, run a chalk line down the wall, then apply the glue along the line.

Attach the first strip in the extreme corner, the second one 24 inches from the corner, and the third at the 4-foot mark.

If your wall is concrete block, you will need to glue the furring strips in place so the adhesive will fill the porous surface of the blocks and bond the furring strip in place. You might also use masonry nails for added support.

If the wall surface is drywall or paneling, you can either nail or glue the furring strips to the wall. When the vertical furring strips are in place, also add horizontal strips at the ceiling, floor, and midway between floor and ceiling. You can add strips at the one-third and two-thirds points for even more support.

With the furring strips installed, you can now nail the paneling to them. Some construction authorities advise leaving a "breathing" space between each sheet of paneling and paneling plank so that when the house expands and contracts as a result of changing temperatures, there will be room for the planks and panels to adjust to the change. The suggested gap between units is the thickness of a dime. Other builders argue that the gap is not important and that the wall behind the paneling will be exposed, particularly when the expansion is at its greatest.

Diagonal installation

Because you have a great selection of lengths when using planks for paneling, you can install the boards or planks vertically, horizontally, or diagonally. When installing diagonally, measure the room from the upper corner on one wall to the lower corner on the other end of the wall. You might need to abut boards end-to-end to reach from one corner to the other. Mark the diagonal line across the room and then mark another line from the opposite corners so the two lines will cross in the center of the room.

Cut your first piece to fit in the corner. The piece will have to be a short triangle so that one corner can fit into the corner of the wall. Do not install the piece yet.

Drop lines from the first diagonal to the floor and to the adjacent wall. These lines should be 6 inches from the line from the diagonal line to the corner. Mark another line from the point where the top line you just made connects with the adjacent wall to where the second line connected with the floor. This last line marks the perfect diagonal course line for all planks.

Now install the triangular piece temporarily and place the second plank segment against the triangle. The upper edge of the second piece should be perfectly parallel to the line that marks the diagonal course. If it is not, make the necessary adjustments. If it is, proceed to nail or glue the triangle in its permanent position. Continue to add subsequent planks until the entire wall is complete.

Your major problem will be cutting the angles correctly for the fit against the adjacent wall and the floor. One simple manner of handling the problem is to experiment until you have the proper angle on one plank end and then use the piece as a pattern for all subsequent cuts. The same angle should be correct for adjacent wall and floor angles, because you dissected the room with geometric accuracy and achieved similar angles.

You can also use your framing square to mark the cut lines. Lay the plank on a work surface and place the square on the plank so that the heel points toward you, the tongue extends up the line for the adjacent wall, and the blade extends along the floor line mark. Keep the tongue near the end of the plank so that you avoid waste. Be sure that the bottom edge of the tongue and the blade are on the same mark with reference to the plank. If the tongue edge is on the 4-inch line where it crosses the plank, then the blade 4-inch mark should be where the blade crosses the plank.

Mark along the outside edge of the square blade and tongue. Cut along the mark. Your angles should be accurate for both wall and floor points.

Length of cuts is a more difficult problem. You cannot hold the board or plank up to the wall and mark it because of the corner and the floor. You will have to establish a pattern to follow for all cuts.

Assume your boards are $3^1/2$ inches wide. Your triangle for the corner will have the following dimensions: $7^1/2$ inches along the top side, $5^3/4$ inches along the floor side, and $4^1/2$ inches along the wall side. The second plank will be $7^1/2$ inches longer along the top side. The other dimensions will remain the same. Each succeeding plank will be $7^1/2$ inches longer than the previous. Make certain to measure and mark along the top edge of each plank.

An interesting effect can be created by installing the planks or boards in a mixture of horizontal, vertical, and diagonal patterns. You have a virtually unlimited opportunity to experiment and achieve the most pleasing designs for your house and your tastes (FIG. 2-6).

For example, you might want to start at the lower right corner of the room or wall and install boards diagonally to the halfway point, then finish the room with the opposite diagonal arrangement. A nice effect can be created by doing the extreme left and right parts of the wall in vertical installation with diagonal arrangement in the center. Or, start the same way and use a V patterned diagonal plan in the center part of the room. Sketch several possibilities to see what your best bets are. When you sketch, use large sheets of paper and make a scale drawing so that you can get a true perspective of the sizes and ratios of the boards.

WALLPAPER

Until fairly recently, hanging wallpaper was a difficult procedure. Now, however, new products have made it easy for the beginner to do a virtually professional job. The new papers are already gummed or pasted, and all you need to do is moisten the gum or paste. Some types of wallpaper do not even require moistening; they can be hung as they are.

Preliminaries

As with all wall coverings, the important task at the beginning is to start with a square corner. If the corner is not square, you will need to modify

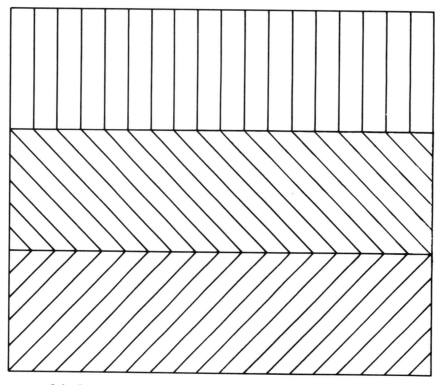

2-6   One of dozens of designs that can be created with paneling boards.

the wall covering so that it will fit neatly and professionally into the corner and leave you the outer edge of the wall covering at a perfect vertical
level.

You can trim wallpaper with scissors, razor blade, or any number of
common household instruments. Always start at the top and work your
way down. When you apply the wallpaper to the wall surface, take care to
get the corners applied perfectly. Line up one corner with the edge of the
previous width and align the top edge with the ceiling. If you cannot do
both because the ceiling is uneven, keep the vertical position true and
trim the ceiling end of the paper. When you reach windows and doors,
you can trim to the very edge of the frames, but it is much easier if you
install the wall covering before the window and door frames are installed.

Before you begin work, clean the wall surface of any dust, dirt, or
foreign particles. If drywall joints are not sanded smooth, now is the time
to take care of them. Rough joints will show badly through the wallpaper
and produce a very disappointing effect.

At wall switches and other wall interruptions, bring the paper down
to the switch level and then slowly work the paper down so that the
switch box is half covered by the paper. At this point use a sharp blade to
trim the paper neatly and you will have a professional-looking job.

Whenever possible, remove switch box covers when hanging wallpaper. Reinstall the switchbox cover after the paper is hung and any edges

that are unattractive will be concealed, just as with window and door frames.

After the paper is hung, use a roller to press out any air bubbles or other irregularities. Start at the center and work your way slowly toward the nearest edges. Do not run the roller over wrinkles or loose paper. This will make a permanent crease in the middle of a wall, a defect that is virtually impossible to correct.

If you are working with paper that must be moistened or pasted, first cover the floor with newspaper or similar protection. Gather the necessary number of containers large enough to accommodate the paper and tools you will be using. You will need a large brush, a roller, and a good cutting instrument, as well as a step ladder and scissors.

Vinyl products

There are many vinyl products on the market today, and these have a great number of advantages over the old-fashioned wallpapers. Modern wall coverings in this category are strippable. This means that if you need to remove the strip or part of it, you can use a thin blade to work one corner loose, and then you can pull the covering from the wall. There is little if any danger of ripping or separating. The new coverings are also washable, a big advantage over the old papers, which had to be replaced when they became dirty. Modern products are also able to withstand sunlight, while older papers tended to fade, causing three or four levels of contrast on one wall.

When installing vinyl products, you will need to moisten the back of the covering thoroughly before installation. If you are using a patterned covering, take great care to match the patterns as you install each new roll. Newer products come with the salvage or waste already trimmed off, so you do not need to make any type of alterations on the edges of the covering. The pattern itself extends completely to the edge and makes matching much easier.

Calculating needs

First measure the room you are to cover from floor to ceiling. In most cases this will be 8 feet. Then measure the complete circumference of the room. Start in one corner and measure along the wall to the next corner. Continue to work your way around the room. Add the total feet on all four walls. Do not deduct the square footage of windows and doors from the total wall expanse. After you have determined the number of linear feet in all four walls, multiply that figure by the number of feet in the height of the room. Divide this number by 60, which is the number of square feet in the typical bolt of wall covering. The result will be the number of bolts you will need. In case of fractions, step up to the highest whole number. You might be able to use fragments of a bolt to fill above and below windows or above doors, particularly if you are using a covering that is not patterned.

Suggestions for decorating and installation

Nearly all wall coverings come with detailed instructions on how to install them. Read the instructions carefully and follow the directions explicitly.

One decorating suggestion is to combine patterns as you did with plank or board wall coverings. Try wainscoting: Install boards or paneling over the bottom 4 feet of the room, add the wainscoting divider—usually some type of wide or thick molding—then install wallpaper or vinyl up to the ceiling.

Another suggestion is to install solid or patternless vinyl on the lower portion of the room and a blending pattern above the divider molding. Some people have found that covering one wall in a room with a vinyl pattern that differs from the other three wall coverings can produce a pleasing decoration. Still others enjoy installing patterned wall coverings both above and below the molding divider. You have many decorating possibilities.

Cost of vinyl wall covering generally runs about $8 per bolt, which contains 60 feet of wall covering. Before you make a determination of patterns, ask the dealer for sample swatches that you can take home to study in relation to your own room style and furniture.

OTHER WALL COVERINGS

One of the newest choices in wall coverings is shakes or shingles. These are usually cedar or similar wood cut exactly like the exterior shakes and shingles. They range in lengths from 8 inches up to 4 feet to allow for great variety in arrangement, and they are $3/16$ inch thick. Cost of these shakes runs about $.80 per square foot, so a room that is 20×15 feet will cost between $450 and $500 to cover, depending upon the door and window space. The shakes can be applied with adhesive, staples, or small paneling nails.

Another new wall covering is wood strips. These strips are very thin and about 3 inches wide and up to 4 feet long. They can be installed in an amazing variety of patterns. You can use adhesive to install them. The strips are thin enough that you can cut them with a knife or even with scissors. The cost ranges from $.50 to $.75 per square foot. One attractive aspect of these strips is that they are flexible enough that they can fit gentle contours in wall surfaces.

In recent years the use of carpet on walls has gained popularity throughout America. The argument in favor of this type of wall covering is that the material comes in soft color tones and blends well with most room decorations. Another plus is that the carpet acts as a form of insulation and tends to deaden noise levels. You clean the wall covering with a hand-held vacuum cleaner.

Chapter **3**

Wall coverings for exteriors

*T*he common types of exterior wall coverings have remained unchanged for years: bricks, wood siding, stucco, vinyl siding, wood panels, decorative tiles, and concrete. All of these have their place in home construction and each has its advantages.

CONCRETE

Concrete is one of the simplest types of covering. The major advantage to using concrete is that you can let it serve as both the interior and exterior wall, and so eliminate the need for studding, plating, and capping.

The durability of concrete is beyond question. It is the cost that sometimes dissuades people from using it. Another disadvantage is that poured concrete requires the use of forms.

In the simplest sense, you can think of concrete walls as basement walls that continued to rise. You can use concrete blocks or poured concrete.

Calculating needs

Laying concrete block is hard work, which is why professional masons charge up to $1.25, and higher, per block they lay. Their fee also includes use of a mixer and other tools of their trade, and if they work per block cost includes mixing the mortar and cleaning up afterward.

In order to figure the cost of a wall or all four walls of a house, consider that 12-inch blocks cost about $1.25 each and 8-inch blocks cost $.85, with variations likely from one part of the country to another. Also add the cost of mortar and sand and the fee to the mason for laying the blocks.

If you want a basic house of 1500 interior square feet with ceilings slightly more than 8 feet high before flooring is installed, the total cost of concrete blocks, sand, and mortar will be about $2700 to $3000. The cost of hiring a mason to lay about 1500 blocks brings the total to nearly $5000 for the walls alone.

You might be able to save money if you know where you can buy cement and sand cheaper than at the dealer's. You will need about 7 yards of sand for this amount of masonry work, and you will also need about 50 bags of portland cement or mortar mix. The cost of these items is included in the estimate given above; deduct from the total cost if you can buy the materials cheaper elsewhere.

One word of warning: Buy mortar mix from a reputable dealer. When mortar mix bags are delivered to a job site, the dealer will not allow the customer to return the mix for a refund. The reason is that when bags of mortar mix are left out on the wet ground or exposed to rain or other precipitation, there is a tendency for the mix to begin to harden and become lumpy. When you try to mix the mortar, you will find that the lumps are difficult or impossible to break down, and not fit for use.

It is possible that the dealer who sells the product cheaper than others has bought unwanted bags of mix from a contractor who overbought and wants a refund. You should be aware also that many contractors will intentionally over-purchase rather than have to order a second time and wait for delivery, and even though the customer has paid—or will pay—for the materials, he might return them for a refund.

You have probably been to large stores where in the home improvement sections there are stacks of mortar mix that stay exposed to the weather for months. Pay a little more if you must, but do not attempt to save money by buying a product that cannot be satisfactorily used.

If you lay your own blocks, you can save about $2000. You can learn to lay blocks quickly so that you can do at least passable work. While it is not a simple job, you can handle it efficiently with a little patience and a great deal of work.

Preparations

Refer to Chapter 1 for a discussion of basement excavation and the construction of batter boards. The batter boards are guidelines that are set up on all four corners of the house perimeter so that the mason will know exactly where the block lines must be laid so that the house will be perfectly square.

It is very rare for a house to be precisely square. Even the most carefully constructed homes have a deviation of half an inch or more, but you need to have the guidelines as nearly square as you can get them.

Remember that when the corners are squared, you can double-check your work in two ways. Measure diagonally across the house space from one corner to another and write down the exact measurement, to a fraction of an inch. Then measure the opposite way and record the results for comparison. If the house is square, the two lengths or readings should be identical. If there is a discrepancy of more than 1 inch, you need to find the error and make corrections.

The second way, which tends to support the first method, is to check each corner for squareness. Measure from the corner stake to a point out each wall line, then measure diagonally from the two points to determine if the corner is square.

Once the house is laid out accurately, the footings must be dug and poured. The concrete for footings for a house this size will cost about $500, and you can save money on both counts if you are willing to do the work. Remember that footings must be below the frost line and they must be about 24 inches wide for maximum support. Consult Chapter 1 for details on this work.

When your building materials are delivered, one of your first tasks should be to buy or obtain plastic sheets and cover the sand, the mortar mix, and the blocks. Most companies will provide plastic covers for the mortar mix when they deliver, but it is up to you to cover the blocks and the sand. The reasons for covering these items are simple. When you leave sand exposed to the weather, it can become so wet from rain that when you mix it with the prescribed amounts of water, your mortar will be so thin that it will not support blocks.

You also need to keep blocks covered because a wet block is very difficult to lay. When you plan your day's work and mix mortar accordingly, you don't want to have to wait for the sun to dry out blocks so you can lay them. While you wait your mortar is hardening and getting more difficult to use.

Laying concrete block

When you are ready to lay blocks, mix mortar by placing 5 gallons of water in a mortar box and adding one-third of the required sand, which is eight full shovels. (Note: In many reference materials the sand is measured in terms of cubic feet. Most people have no way of measuring cubic feet; instead think of a cubic foot as a shovelful of sand.) Mix the sand and water thoroughly and then add the contents of one 70-pound bag of mortar mix or portland cement.

Mix these materials until the sand is completely mixed with the mortar mix. Do not be content simply to wet the materials: they must be completely mixed for best results.

Add the remainder of the sand slowly and mix. If you are mixing by hand you will need a hoe and a shovel. Use the hoe to rake through the mixture and use the shovel to turn it. Place the shovel at the end of the mortar box and push it downward so that all of the sand on the bottom is freed and turned to the surface. Do the same along both sides and the other end. Do not stop until the mixture is plastic enough so that it can be worked easily and stiff enough to hold a crease.

To test the mixture, shove the trowel in and turn a trowelful of mortar. Then rake the trowel point through the turned-up mass so that you have a deep crease. If the crease holds, the mixture is thick enough. If it does not hold, add a shovelful or two of sand and mix the new sand completely. If you have to add six or more shovelfuls of sand, you will need to add a shovelful of mortar mix to maintain the necessary strength of the mixture.

A story pole can be used to keep course work on line (see Chapter 1). Mark a pole or board for use with blocks. A block with its mortar joint is 8 inches high. Make a mark every 8 inches on the board. You can mark four boards and install these at the corners of the house if you wish. The boards can be used later, so there is no waste.

Attach lengths of boards or 2 × 4s to the story pole about 5 feet from the bottom and run these boards out to the sides of the footings. Drive a stake at the end of each support timber and drive a nail through the stake and into the timber. When you set up the story pole, place it with one end on the footing and use a level to position the pole in a vertical manner. Hold it in place until both support timbers are nailed to it. The support timbers should run parallel to the footings and all four story poles should be perfectly vertical and so well supported that they will not give or sway off a plumb position.

Run a block line from the first 8-inch mark to the one on the other corner that corresponds and pull the line tight. You might find that if your footings are not level the line will be higher at some points than at others. This means that you will need a deeper mortar bed in places.

As soon as the line is tight, you can begin to lay the mortar bed. First, if you have not already done so, clean the footings. Shovel away all mud or dirt that might have washed in and sweep the sand and other accumulations off the concrete. Then scoop up full trowels of mortar and spread it across the footings. Do not lay a bead or row. Completely cover the footing to a depth of 2 inches or so.

While the mortar is hardening slightly, you can set up lines for other walls. If you allow the mortar to set for about 10 minutes, you will find that the heavy blocks will not sink so deeply into the mortar. If you do not allow settling time, the blocks might sink so low that they will not be in line with the block line and your joints and courses will not be accurate.

Before you start to do any actual block laying, you can either measure or set blocks along the line to see if you will end the course with a line of full blocks. You do not want to start trying to cut blocks or shape them in order to fit them into the course unless you are experienced as a mason and unless there is no alternative.

When you measure, remember that a block plus a mortar joint is 16 inches long. If your longest wall is 52 feet long, you will have a total of 624 inches in that wall. A block is 16 inches long, and if you divide 16 into 624 you will find that you can lay 39 blocks without cutting one and your course will be nearly filled. If the wall is 32 feet wide, you will have 384 inches. Divide 16 into 384 and you find that you can lay 24 whole blocks on the short walls.

You now see one of the advantages of choosing house dimensions that are divisible by four. If you have to cut or shape blocks, you not only spend a considerable amount of time in this work but you also damage many blocks beyond use.

When you start laying you might wish to measure down the wall line to a point 17 feet and 4 inches. Mark the point on the footing beside the mortar bed. Then move from that mark another 17 feet and 4 inches. These marks represent the one-third points along the wall line. At the first mark you should have laid 13 blocks, at the second mark another 13 blocks.

If you find that you passed the mark, even by a fraction of an inch, before you had laid 13 blocks, your mortar joints are too wide and you will need to tighten them in order to get back on your schedule. This is a very important consideration. Every course you lay in the entire wall should be as uniform as the very first one. If you do not maintain a rigid spacing, your mortar joints will not match later and your bond will not be correct.

If you have laid 13 blocks before you reach the mark, your mortar joints are too tight and you need to widen them. The resulting problem is that you will now have to space blocks unusually wide and pack the joints with mortar. Extremely wide mortar joints might be in violation of building codes in your area and, if so, you could be asked to take down the wall and rebuild it.

If your mortar joints are too wide, go back several blocks and push steadily until the block moves slightly. Then move to the next block, and keep repeating this process until you have created a wider space. Reverse the process if the joints are too thin.

When you start to lay blocks, notice that in the center of the block there is a partition which is wider on one side than on the other. This space enables you to grasp the block better while you are lifting and positioning it.

To mortar or "butter" a block, stand the block on end and scoop up a small load of mortar on your trowel. Hold the trowel with the handle far back in your hand and your thumb on top of the handle and extended down the handle as far as it will reach comfortably. Hold the trowel so that the blade is parallel to the ground. When you need to load the trowel, maintain the proper grip and turn your wrist and slice the trowel into the mortar.

As you approach the end of the block, turn your wrist so that the trowel is turned until the blade is parallel with the side of the block. Start with the point of the trowel dipping to a four o'clock position and slice the trowel downward so that the blade skims past the side of the block and the mortar is spread in a neat bead. Do the same on the other side of the block.

Before you lift the block, lay the back of the trowel blade on top of the mortar bead and press the mortar sideways and into the groove at the end of the block. This pressure will keep the mortar from sliding off the block when you are positioning it.

When you are ready to lay the block, lift it by the middle partition and the end that is facing upwards—where the mortar is spread. Hold the block with one hand in the middle partition and the other hand in the indentation at the end of the block.

The first block you lay in the course will be an end block. It will have one square end and one irregular end. Lay it so that the square or flat end is to the outside. You will want to use an end block on each end of all wall courses.

On subsequent blocks, stand the block on end, butter it, and when you lift the block hold it horizontally and set it in the mortar bed so that the buttered end of the block is jammed into the end of the previous block. Press the block firmly into the one it joins so the mortar will be pressed into the grooves on the end. This is what creates most of the holding power of the mortar.

When the first course of the wall is laid, you will need to lay at least one block on the short walls so you can start the bonding process. Lay a mortar bed and when the mortar is stiff enough, butter a block and abut it to the side of the first block in the course you just finished.

Each time you lay a block, check that it is level and square. First check to see if the block is correctly positioned with reference to the block line. If it is not, tap it with the end of the trowel handle until it is situated properly.

Lay the level across the block to determine whether it is level from side to side. If it is not, place your hand with gentle but firm pressure on the high side and tap the edge of the block with the blade of the trowel turned vertically. A gentle tap will cause the high side to sink into the mortar enough to give a level reading. Lay the level next on the long surface of the block to check for a level reading. Correct any improper readings as you did earlier.

As you progress in your course work, you will also need to place the level against the side of the wall with the level position in an up-and-down manner to be sure that the wall is perfectly plumb. As you move along the course, hold the level occasionally against the sides of the blocks so that the level is horizontal. Check to see if the sides of all blocks are straight and are touching the level at all points.

When the first block is laid on the short wall, you are ready to lay the first block on the second course of the long wall. Note that the first block you laid extends its full length along the course line. The first block of the second course must be set so that it crosses the first half of the block. Half of the block you are now laying will rest on the first block in the course and half of it will rest on the first block in the second wall. This is called bonding. When the courses overlap in this fashion at no point will there be two adjoining mortar joints that are aligned.

As you proceed with subsequent courses the vertical or head joint of all blocks in every other course should match precisely. For example, the joints of block nine in the second course should align with block nine in the fourth course.

Any time you notice that the bond joint in the blocks is not aligning properly, you need to tighten or widen mortar joints accordingly. If you do not, you will need to perform what masons refer to as "clipping the ears" of the last block in the course. This requires laying the block on its side and using a mason's hammer chip away the grooves on all four surfaces of the block until you have shortened it by 1/2 inch or more. Sometimes even the clipping might not be enough, and you will have to break the block so that the end is removed. This process is not easy, and often results in a broken block that has no real use except as filler later on.

To break the end of a block out, first lay the block on top of another block so that the solid surfaces are together. Use the mortar hammer to start tapping a line across the block just behind the solid end portion of the block. Hold the block with one hand so that your fingers are inside the end of the block. When the block starts to crack, you will hear the noise of the hammer change to a deeper sound and you can also feel the cracking begin slightly. Stop at this point and carefully turn the block so that the opposite flat surface is facing upward. Tap along the corresponding line you used before and you will feel the block again start to crack. Keep tapping until the end of the block can be separated from the remainder of it.

You may now butter the broken end of the block and install the block into the course. You will, depending upon the number of blocks in each course and the distance to be covered, find that in alternate courses you will need a half block. When you order blocks, ask the dealer to send you a supply of halves and grooved blocks. With halves you won't break so many blocks, and the grooved blocks can be used for window installation.

Installing window block framing

Determine where your windows will be installed, but be prepared to make slight modifications. You might have decided that you want the windows 9 feet from the corner and 5 feet apart midway between corners and door. When it is time to install the windows you might find that the distance you have selected requires breaking and shaping a block in each course to make the distance work out correctly. Check if you can—by moving the window location 3 or 4 inches or even slightly more—cause the blocks to work out to an entire block rather than to have a shaped block in the course. You will save time and money on broken blocks by doing so.

If you have chosen aluminum windows for your walls, you will find on the edge of the side window framing a series of tiny grooves that are similar to a perforated line on paper where it is to be separated. You can alter the width of your window slightly by removing the grooved section or sections. Do not make any changes until you have laid the grooved blocks for the window opening.

Determine the height of the windows and the height needed to set off the window opening. If the window opening is to be 40 inches wide, mark the point where the 40-inch space will be located, and on each side of the window lay a grooved block with the end of the block corresponding to the marks denoting the window opening.

Note that some blocks have a vertical groove running down the center of one end. This groove is designed to allow the framing of the window to fit into it so that the window can be moved slightly to modify height. The end block in every course on both sides of the window should be grooved, and the grooves must be aligned perfectly (FIG. 3-1).

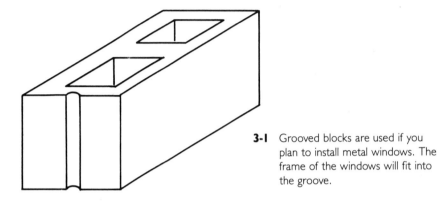

3-1   Grooved blocks are used if you plan to install metal windows. The frame of the windows will fit into the groove.

When you have laid two or three of the grooved blocks on each side, slide the window frame into the grooves and lower the window into place. The fit will not be tight, and the window can be moved freely. Later you can use caulk to seal the window in the frame.

Keep laying grooved blocks in each course until you have reached the maximum height needed for the window. You can continue to end each course with grooved blocks if you prefer. Remove the windows to keep them from being marred or broken, and when you need them installed simply slide the frame into the grooves and slide the window into position.

The window frame might be too wide for a good fit into the grooves. If so, grasp the edge of the scored strip with a pair of pliers and bend back and forth until the top of the strip breaks free. Grasp the end of the strip in the pliers and start to roll the pliers down the strip. As you do so, the strip will curl around the pliers and you will quickly have the entire strip removed. The window frame will then be narrowed by 1/4 inch.

Remove the strip on both sides if necessary. You will see that there are two or three more scored strips on each side of the window, and you can remove as many of them as needed for the best possible fit.

When you reach the top of the window opening, you have a choice of installing lintels or continuing the rough opening to the top of the wall line. In ancient buildings the lintel was a huge beam installed over the window or door openings of a structure. In modern buildings the lintel can be a length of angle iron or similar metal that will support the weight of the building above it. You can buy angle iron for this purpose. It is thin enough that it will not interfere with the mortar bed joint line or block line and it will permit you to lay blocks across the top of the window opening.

If you choose to omit the lintel, you can install a header above the window opening, which will help support the weight of the building above it.

Do not be concerned about the open space above the window. In addition to the header there will be a sill installed and the combined strength of these two supports will be more than adequate for the weight above the window.

(See TAB Book #3435, *Rough Carpentry Handbook,* for a discussion of header construction.)

Framing rough door openings

When you have marked your rough door opening, you must end each course with an end block so that the opening will be uniformly straight. These end blocks will be installed on both sides of the door opening.

As you lay these sections of your walls, take great care to keep the plumb line of the end blocks constant. Use the level regularly, and occasionally measure with a tape from one surface to the other in the door opening.

As with the window opening, you can continue the door opening lines to the top of the wall, or you can install a lintel. Remember that if you omit the lintel you will need a header and sill timber for support above the door.

Buy the frame for your door at the same time you purchase the door and install them together. If you remove the door from the frame the block alignment might not be exact, causing the frame to be pushed in slightly and preventing the door from fitting into the opening when you are ready to install it.

Stand the door and frame in the rough opening. Attach brace supports to hold the door and frame steady while you lay the blocks up to the frame and connect the frame to the wall. The door and frame should be set in place as soon as the rough door opening is established.

In each course set an end block against the door frame. Take care to keep mortar joints consistent so there will not be any joints too wide for proper strength or too narrow for mortar. The door opening and the window openings will weaken the wall somewhat and you want to be particularly careful to maintain strength in every possible area.

When you rise from one course to another, you solidify the door frame by positioning two or three mortar nails per course so that they are slanted nearly flat. Place these nails on the block that abuts the door frame and drive the nails into the frame an inch or slightly more. The nails should be deep enough into the timbers to hold well, but the points should not show through to the inside of the framing. When you lay the bed joint for the next course, the mortar will surround the nail head and part of the shank, and the next block will be placed upon the mortared nails. When the mortar sets up and cures the nails will be firmly secured, and the door frame will be held in place tightly and securely (FIG. 3-2). Install these nails in every other course and on both sides of the door frame.

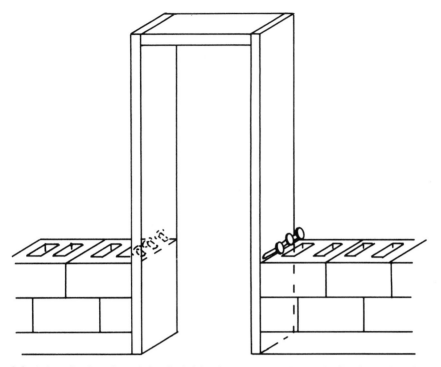

3-2 When the door frame is installed, drive three mortar nails into the framing timbers for each side.

When you reach the top course for the wall, exert great care to achieve a level and true course at the top. Double-check for any irregularities. Your rafters will rest on this top wall, and you do not want one rafter higher or lower than the others. Such a small discrepancy can create difficulties when you are establishing a roof line.

One item that is required for block building in many areas is wire reinforcement. Sold under such brand names as Durowire, these heavy wire products are 10 feet long, 8 or 12 inches wide, and thin enough to be installed atop a course of blocks and then covered with mortar.

Install reinforcing wire by starting at the end of a course and laying the wire framework so that the end of the first unit is 2 or 3 inches from the end block. The next unit should be placed so that it overlaps the other end of the wire by 5 or 6 inches. Do not stack the wire units one on top of the other. Lay them out so that the end wire sections fit beside those just placed. When the entire wall is covered, lay a mortar bed and then lay blocks across the wire and mortar.

Building codes often require one layer of reinforcing wire in every third or fourth course of blocks. Check with your local code before proceeding.

BRICK FACING

Follow the same basic approach in laying bricks as you did in laying blocks. The major differences lie in the weight of the masonry materials and the space covered. One 8-inch block will occupy 1 square foot of wall space, while several bricks are needed to fill 1 square foot of wall area.

Another difference between brick facing and block walls is that the blocks form the entire wall, while in brick facing you have a traditional wall frame, complete with studding and sole plate and cap plate. The sheathing is installed as if the wall would be covered with clapboards or siding. The bricks are then laid on top of the block masonry that formed the foundation wall and up the side of the sheathing until the top of the wall is reached.

Many builders use 12-inch blocks for all of the underground masonry walls, and at the top of the foundation wall use a course of 8-inch blocks that are plumb on the inside rather than the outside of the wall. There will be a 4-inch recess, and this is where the brick facing will be started.

You will use block line, level, and all of the other tools used for laying blocks. The mortar beds will not need to be as deep and the trowel can be much smaller than the large trowel used for blocks.

For various reasons, masons suggest that bricks be sprayed with water before they are laid: A dry brick absorbs water from the mortar and reduces the bonding ability; the water washes away dust and dirt that will interfere with bonding; and mortar spreads better under a brick that has a wet surface.

A process called *hydration* allows mortar made of portland cement to harden properly, and moisture is required for this process. If the bricks soak up the water needed for hydration, not enough hardening will occur, and the mortar will be crumbly.

The procedure for rough window and door openings is similar to block masonry. One difference is that with brick work you will need to install bricks at the sill openings of windows, while in block masonry you fill the space with concrete.

Keep in mind that bricklaying is a complicated process requiring years to master. Do not expect to do expert bricklaying in a matter of hours. What you can do is learn to lay bricks so that your work is presentable and pleasing to the eye as well as structurally sound.

As you are laying the stretcher bricks, lay the mortar bed for the entire course and then lay all the bricks at one time. You do not need to worry that the mortar will dry before the bricks are laid unless the courses are abnormally long, or you are unusually slow in your work, or there are interruptions. When there are interruptions, cover the mortar in the mortar box using a sheet of plywood or plastic. Mortar on the mortar-board can be dumped back in the box where it will be kept sufficiently moist until you are ready to use it.

STUCCO

In Chapter 1, stucco was described as a method of waterproofing concrete blocks. You might want to consider using stucco to cover the entire exterior wall space of your house. Stucco and plaster are essentially the same product. When it is applied indoors, the term used is plaster; when used outdoors, it is stucco.

An explanation of terms

The key terms used in stucco work are base coat, finish coat, binder, aggregate, plasticity, cementicity, and hydration. All of these terms are also applicable to mortar, except for the first two.

The *base coat* of stucco is simply the first coat. It is used to cover the concrete or lath work and to provide a bond between the surface and the stucco. The first coat or base coat is somewhat similar to a primer coat of paint: it prepares the surface for what is coming next. Often the base coat is thinner than the later applications. Thin stucco simply by adding water and mixing until you have a more plastic state.

The *finish coat* is the final application. This coat will be smoothed and finished until it is both pleasing to the eye and fully practical.

The *aggregate* is whatever is mixed with the portland cement and water. The most common aggregate is sand, but it must be mason's sand, not common sand. The problem with common sand is that it usually contains a lot of foreign matter, small pebbles being the most problematic. These pebbles will not permit you to smooth the stucco well, and you will be forced to remove the pebbles by hand before you can continue work.

Plasticity is the workable nature of the stucco or mortar. If the stucco is too stiff, it will go onto the walls thickly and create waves, uneven surfaces, and unsightly humps. It can also crumble or crack. After it has cured for several days, too-dry stucco will crumble simply by rubbing it firmly with your finger. If the mixture is too thin, it will apply easily. The problem is that it will not cover the surface and render the lath work or concrete block invisible. The aggregate will also tend to separate from the mixture as well.

Cementicity refers to the ability of one matter to adhere or cling to another. Usually cement provides the adhesive quality needed for concrete products.

The *binders* in a mortar or plaster mix are the ingredients that hold the combination together. Gypsum, lime, and portland cement are common binders.

As mentioned earlier, hydration is the process through which the mortar or plaster binds itself to the stone, brick, concrete, or wall surface.

Some of the major advantages of stucco are that the finished wall has most of the properties of concrete: stucco is hard, fire resistant, waterproof, will not crack after repeated wetting and drying cycles, resists severe weather changes, will not wash, and resists rot and fungus growths.

When you are mixing the stucco, you can use portland cement Type I, water, and sand. Again, be sure that the sand is as free as possible of organic matter, clay, loam, or vegetable matter. These impurities will cause the stucco to crumble later.

Applying stucco

When you are applying stucco—as opposed to plaster—the primary tools used are a regular mason's trowel, a device called a hawk, a rectangular trowel, and a float. The *hawk* is a square sheet of thin metal or flat wood such as plywood with a central handle. The hawk carries mortar from the mortarboard to the place where the mortar is to be applied. You can hold the hawk in one hand, with your fingers grasping the handle and the flat metal or wood sheet resting upon the circle formed by your index finger and thumb. With your other hand you load your trowel and apply the mortar. The primary advantage of the hawk is that it saves you considerable time and effort because you do not have to bend to reach the mortarboard and you do not have to make so many trips from your position to the mortarboard.

The rectangular trowel is a thin piece of flat metal with the handle fastened to the center of the back. The major uses of this type of trowel are to smooth surfaces and to apply mortar or stucco with the trowel to a flat surface (FIG. 3-3).

The *float* is used to slide over the surface of the mortar or plaster and to fill in any holes, cracks, or thin spots in the work surface.

When you apply the base coat of stucco, hold the trowel in the usual manner, but when you scoop up stucco turn the trowel so that it is upside down. When you dip up the mortar, it will be on the bottom of the trowel, but now the bottom is facing upward. As you lay the stucco on the surface, make a sweeping upward motion and smear the stucco as uniformly and as thinly as you can while still covering the surface. Work rapidly, smearing in a series of quick motions that curve upward (FIG. 3-4).

When you have covered an expanse of roughly 30 square feet, pause to smooth the stucco that has already been applied. Do this by dipping your trowel in a bucket of water and, while the trowel is dripping wet, use the back of it to flatten and smooth the mortar. You will need to make several passes over the surface before you have achieved a satisfactory smoothness.

Use metal reinforcement when applying stucco to any surface that does not provide a good bond. Wood, vinyl, and similar products often require metal.

When applying stucco to wood frame buildings, you might need to install a wire base to support the stucco. Examine your local building codes carefully before you attempt to apply stucco over wood or vinyl or metal surfaces.

You can apply stucco directly to cement or concrete block masonry. The base coat, as you have already noted, should be thin—no more than 3/8 inch. Let the base coat cure until it is dry enough and strong enough to

3-3 Smoothing any type of concrete or mortar work requires patient and careful use of bull floats and trowels. This technique works equally well on a patio or a wall.

support the weight of the second coat or finish coat. Usually 4 to 6 hours will be required for curing.

When you apply the finish coat, thin the coat by applying enough pressure so the mortar is slightly compressed. You can leave the first coat

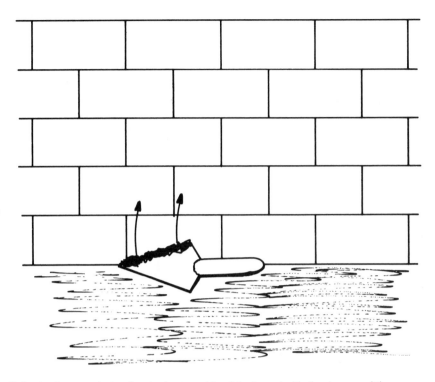

3-4 Apply stucco by loading the bottom side of the trowel with the stucco and then pressing it into the wall surface with a sweeping upward motion. Let each stroke lap the previous one slightly.

rough if you wish. The rougher the surface, the better the second coat will adhere or bond to the first.

You can buy a pigment mix that can be used with most stucco mixtures. This mix will color the stucco and make it more suitable to the color of the rest of the house. Ask your dealer what pigments are available.

WOOD SIDING

Most board sidings are referred to as clapboard siding. These exterior wall coverings consist of boards that are usually 4 or 5 inches wide and 1 inch thick or less. You can buy siding lumber that has two square edges, or that has a beveled edge.

To install wood siding, you will need a carpenter's hammer, level, square, a story pole, and common nails. Ask your dealer to suggest nail sizes that will work best with the lumber you are using.

Using a story pole

The story pole serves the same purpose here as it did in the masonry courses. On the masonry story pole you included in the measurement the height of the masonry block or brick as well as the thickness of the mortar joint. With the siding story pole you use only the width of the board, less the amount of lap. You can use a regular length of siding stock for your story pole if you wish. If you are using 5-inch boards with a 1-inch lap, then mark the levels every 4 inches.

When using the story pole, start the pole or board even with the bottom of the first board you install. Do not measure from the ground because the ground level is not likely to be the same at all corners.

Installation

Your first step in installing siding boards will be to install the corner boards. These are usually two boards, one of them 4 inches wide and the other 5 inches wide, on the corners of the wall you are planning to cover. Nail the 4-inch board to the corner post of your wall frame after first lining up the board with the outside edge of the corner post. Then install the 5-inch board so that its outside edge is flush with the edge of the 4-inch board you have just installed.

The reason for using boards that are not the same is that you will want the corner boards to look symmetrical, and when you lap the thickness of one board you will have a corner that looks poorly balanced. The added inch of width on the one board offsets this (FIG. 3-5).

Before you install your first siding, nail a thin strip of wood (2 inches

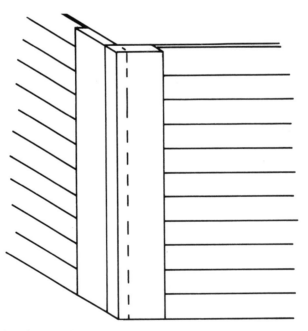

3-5   Corner boards cover the ends of siding timbers or units and give the corner a neat, finished look. The outside edge of the first board is flush with the outside edge of the siding. The second board laps the outside edge of the first board and fits against the siding.

wide and 1/2 inch thick) to the foundation joists just above the foundation wall. The purpose of this strip is to make the bottom of each board extend from the wall slightly so that rain will not run directly down the siding and onto the foundation wall. Prolonged exposure to moisture can cause discoloring and other unsightly defects.

When the starter strip is installed, align the bottom of the first board so that it extends slightly below the point where the joists meet the foundation wall. If you leave this crack exposed, you will permit unwanted moisture and insects to enter the house.

With the board in place, sink the nails through the board and into the framing timbers. Drive the nails in about 6 inches from the ends of the board and 1 inch from the top. This will keep the nails from showing and from being exposed to the weather and rusting.

When you have driven the first nail, start a nail at the other end of the board. Then hold the level on top of the board to get the correct position before you drive the nail the rest of the way. Drive a nail every 3 feet across the top of the board.

When you are ready for the second and all subsequent boards, there are two methods of handling the task. First, use a measuring tape and chalkline and locate and mark a point 1 inch below the top edge of the second board. Do this at several points and drive a nail an inch or so deep into the wood along the marks. Set your next board in position so that it rests upon the nails. Your board will be firmly held by the holding nails while you drive the permanent nails to install the board. Again, drive the nails 1 inch from the top so that the lap from the third board will cover the nail heads. When the board is nailed securely, the holding nails can be removed (FIG. 3-6). Use your level frequently to see that you are maintaining the proper level.

The second method is equally simple. Start a nail in the second board and hold the board in place while you sink the first nail. Then use the level to secure a perfect position for the board before you nail at the opposite end. When you have two nails holding the board, use the story pole to determine if you are keeping the boards properly aligned. Continue the installation pattern up the entire wall of the house. Do other walls in the same fashion.

All windows and doors and other wall interruptions should be framed before you install siding. Frame windows by installing framing boards on both sides and then above the windows. Do the same with doors but install framing boards only on the sides and top.

If for some reason your window or corner or door framing boards are not installed with a true vertical positioning, your siding boards with perfectly squared ends will not fit against the boards evenly. If you cannot reasonably correct the problem, hold the siding boards for all subsequent cuts so that the siding board extends slightly past the edge of the framing board. Make a pencil mark where the framing board and the siding board meet. Mark both top and bottom of the siding board and then use your square to mark the cut line. Cut along the line and the board should fit well.

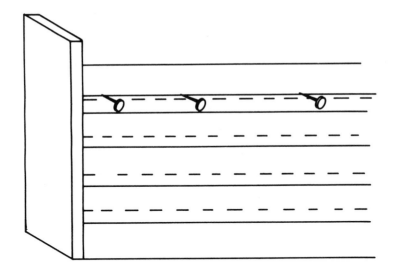

3-6   Use a chalk line to mark the lap distance on the siding board. Drive in two or three nails to hold the next board or unit in place, and when the board is nailed securely the holding nails can be removed.

VINYL SIDING

Wood siding is one of the most attractive exterior wall coverings available, but it has several disadvantages. First, the wood can decay if it is not weatherproofed and painted or stained. Also, wood can crack, warp, or bow. A third problem is the effort and expenses—and sometimes danger—involved with keeping wood painted. With these problems in mind, if you are looking for an attractive siding that will never need painting and will not warp, bend, bow, or crack under normal circumstances, investigate vinyl siding.

For years vinyl was a distant second to aluminum siding, but steady improvements in vinyl products have caused the vinyl siding to gain rapidly in popularity. One of these improvements is that vinyl is colored all the way through, not just on the surface, so the paint cannot peel, chip, or flake. Once vinyl is properly installed, the surface will never need maintenance, except for an occasional washing or cleaning. Neither will vinyl warp or crack under normal conditions.

Like any building material, vinyl works best when installed according to specifications. If you nail too near the end of a section, the lap ends can fit poorly, and if you nail too tight you can create unsightly problems. You need to have a solid and even nailing surface behind the vinyl as you work.

For years vinyl was available in a limited number of colors. Today the color range is very wide and you have a much greater selection.

Insulation properties also caused many people to shy away from vinyl and aluminum in the past. However, the thicker vinyl materials, coupled with the use of furring strips to create air space behind the siding, has enabled vinyl users to have one of the finest insulation ratings available in the field of thin or sheet or hardboard sidings.

To install vinyl siding you will need, for the most part, only the typical tools used in other forms of carpentry. Other tools you might find necessary include a lug punch and lug key. These tools are known by a variety of names. Ask for the tools that will allow you to unlock vinyl panels already in place and the salesman will be able to assist you.

Calculating needs

Your first step is to determine how much siding you will need for the job. Ask the salesman whether you can return any materials not used or damaged. In many cases you will be able to return any unopened or unmarked materials, but not all stores offer this consideration.

When you are calculating the amount of siding to buy, figure the square footage of the complete exterior of the house, including only that part of the house to be covered. Do not include foundation walls or similar areas. Take one wall at a time. Measure the wall from corner to corner and from the top of the foundation wall to the eaves. Do your calculations in feet and multiply the length times the height of the wall in order to learn the square footage.

Where there are peaks or gables, if the peak is fairly typical, you can measure from the widest point and then from the center of the widest point to the peak. Assume that the gable or peak is 30 feet across and 12 feet from the center of the wall to the peak. You will have 360 square feet to cover. You have only half a rectangle rather than the entire rectangle, so you can then take half that amount, which is 180. Or, you can count the entire amount and then not measure and count the gable or peak on the opposite end of the house.

When you have measured all the walls, including recesses and alcoves, add the square footage for each surface to be covered. Then measure the square footage of all doors and windows, and add them together. Subtract this amount from the complete square footage of the exterior of the house. Then add 10 percent to the final figure.

Assume that you have a total of 1600 square feet of exterior wall space, and of that amount 350 is window and door space. Subtract the 350 from 1600 and you will have 1250 square feet. Then add 10 percent to the total. You will have 125 square feet to add to the 1250, giving you a total of 1375 square feet. The extra 10 percent allows for any waste that occurs. You already know that it is virtually impossible to use every square foot of wood you buy; the same is true of vinyl or aluminum siding.

You will also need a good ladder system or sets of scaffolds from which to work. You can rent or buy the scaffolding. Unless you will have frequent use for the scaffolding later, it will be better to rent the equipment.

When you set up the scaffolds, take care that the bucks rest upon a flat and solid surface. If you are working over soft dirt, the legs or feet of the bucks will sink slightly and create an unsafe work area. Use boards or similar surfaces under the legs to prevent sinking.

When the bucks are set up, run thick boards across the frames at the best working height for you. These boards should be 2 × 6 or larger and free of cracks or other weaknesses. Allow yourself a work area at least 4 feet wide. Spread a panel of plywood over the boards to give yourself a smooth and even footing.

Before you buy the siding, decide whether you wish to use vinyl for the fascia and for guttering. You might be able to find these in vinyl, but you might also have to settle for aluminum. Figure the fascia in linear feet rather than in square feet.

Installation

After you have purchased the siding materials, nail up any insulation and furring strips that are necessary. If you choose to install vinyl over the existing wood siding, you will need furring strips for nailers.

When you are ready to work, use a chalk line to mark for the starter strip. This strip should be located where the foundation wall meets the existing siding or insulation. Use aluminum nails or rust-resistant nails for all vinyl siding work, and be sure that the nails sink slightly less than 1 inch into the nailing surface behind the siding. If you are using insulation boards, the nails must penetrate the insulation boards and sink into the wood behind the foam insulation.

Where the starter strip meets the corner boards, leave a small space to allow for the vinyl to contract and expand. You can use a 20-penny nail as a spacer.

Install the corner posts when the starter strip is in place. You will see nailing slots along the edges of the corner posts. Place nails about midway in the slots. Do not drive nails so deep that they crinkle or bend the siding.

When both corner posts are installed for the wall, you will then be able to snap the first siding unit into place. You will feel it lock into the corner posts, and you will also have nailing slots along the upper edge of the unit. Again, drive nails in at the midway points in each slot.

Repeat the process until the wall is covered or until you encounter windows, doors, or other wall irregularities. You might need to add flashing or furring strips, or both, at windows and doors. If there is no flashing already there, you can buy it by the roll at most supply stores. A 50-foot roll of flashing sells for about $20. You will need tin snips to cut the flashing. After you cut it for length, you will need to shape it into a right angle or V so one side can be nailed against the wall and the other edge can extend to prevent water from entering the window or door.

Around the windows you will need to use the J channel units of siding. You might need to nail furring strips around the window to give you a good nailing surface. The vinyl siding will then lock into the J channel units. Do the same with doors and similar wall line interruptions. When you have finished with the wall sections, you can add the finish trim.

Most manufacturers include directions with the siding, and you should follow the directions completely. If the materials do not have complete directions, not only for installation but for care of siding after installation, ask the dealer for literature on the project. If you do not follow the manufacturer's directions carefully, you could negate the warranty on the siding. Just as some roof shingles are not covered by warranty unless felt paper is installed under the shingles, some siding might not be covered by warranty unless spacing directions are followed and proper nails used.

Products are changing constantly, so be sure to obtain the most up-to-date installation procedures for all types of synthetic siding. Check with local building codes for any particular local restrictions that apply.

SIDING SHAKES

Among the most popular exterior wall coverings on the market today are cedar or wood shakes. These are large, thin expanses of wood that come in a variety of sizes. They are installed in much the same manner as roofing shingles.

Installation

The customary manner of installation is to first install the corner and window and door framing, then install the starter strip for watershed effect. You can then use a chalk line to mark the shingle or shake starting line. Make the chalk line 1 inch from the point where the foundation wall meets the joists. If the shakes are 15 inches high, then chalk the line at a point 14 inches above the foundation wall.

Start installation by nailing up the first shake so that the outside edge is set firmly against the edge of the corner framing board. Ask your dealer what type of nails to use. The size will depend largely upon the thickness of the shakes and the type of wood used to make them.

Install the second shake so that the top edge is aligned with the chalk line and the edge next to the first shake is about $1/8$ inch from it. This tiny space allows room for expansion and contraction of the house.

After the first course of shakes is installed, go back and install another layer of shakes on top of the first row, but staggered so that the cracks between shakes will be covered by the second row. Use shakes of different widths so that no two lines will be on top of each other. There should be at least 4 inches lap for each line.

The next row of shakes should lap over the top half of the row below it—placed like the first row. Then each subsequent row will lap halfway down the shakes below and no two cracks will be aligned. By installing in this fashion you eliminate the greatest danger of water leaking or seeping between the shakes and damaging the interior walls.

PLYWOOD SIDING

There are many beautiful and economical plywood sidings available on the market today that will weather properly and can be used on exterior walls.

One of the best points of these materials is that you can cover such a great amount of space in a short time. While plywood siding products are fairly costly—$12 to $20 per panel—the time and effort you save will offset the initial purchase price. The sheets of plywood can be installed easily and with only a hammer, level, and saw for your basic equipment.

On exterior walls you do not need to put up sheathing as you would with ordinary siding. The plywood offers considerable insulation qualities and you need only the plywood and insulation between studs when you finish interior walls.

Installation

Start by aligning the first panel of plywood with the outside edge of the corner post. If you are working alone, use a C clamp to hold the panel while you start the first nails. You can also start installation by driving two nails between the foundation wall and the floor timbers. Then lift the plywood and rest it on the nails. You will need to apply pressure to keep the panel from falling, but you can do so easily while starting the nails.

You need to have a square house when you begin work. If there is a problem, align the first panel so that it is level and shape it to conform to the corner post.

Do not put up window and door framing until after the plywood is installed. You can install the panels right up to the edge of the window itself and then install framing over the plywood.

When you reach a window, measure from the last installed panel to the edge of the window. It is a good idea to measure at the top and at the bottom of the window to be sure there has not been an error in the installation of the window. Turn the plywood face down and measure and mark the distance from the window. Use a circular saw to cut along the mark. If the face of the panel is upward, there will be splinters.

Use one panel of plywood for cutting the sections that will be installed above and below windows and above doors. One panel will be enough to install around one window and one door. You might be able to use one panel around two windows if you mark and cut carefully.

Exterior plywood can be stained or painted, as you prefer. You can also buy prestained panels.

Chapter 4

Doors and windows

*T*wo or three decades ago a window was simply an interruption in a wall designed to let in air and light. Windows were made of wood and were among the cheapest of building materials. Today you can buy wooden windows, as well as those made from aluminum and vinyl that never need painting and last for decades.

Today you will also find that windows can be a very expensive part of building. The best windows can cost you from $450 to $600 each; some speciality windows even more. The cost of windows for a basic house of about 1500 square feet will total about $5000. You can cut the cost by shopping for best buys and by settling for less glamorous windows. Windows in the lower price range cost from $40 to $150. If you choose these windows, you can save more than $2500.

The high cost of windows is not due entirely to inflation. Some modern windows have dual panes with dead air space between them, so you do not need storm windows. The money you save by eliminating storm windows from your building costs can be significant. Storm windows can cost about $60 or slightly less each, which represents a savings of about $1000 on a 1500-square-foot house.

Start your comparative shopping by telephoning several dealers in your area. Read the building supply ads in your newspaper. When you have narrowed your preferences to two or three windows, drive to the supply house and make a personal inspection of the windows. A cheaper window is not necessarily a bargain; it might lack detailed workmanship and quality materials. If you will have to replace the windows within the first 5 years you are in the house, you lose money rather than save it.

You might want to investigate buying more expensive and better-looking windows for the front of the house and less expensive ones for the back of the house. Also examine the guarantees that come with some windows.

Not only have the cost and materials in windows changed, the styles have changed considerably. You can now buy windows that are installed vertically or horizontally. Some of them snap out for easy cleaning. Some have dual panes as well as storm windows built onto one frame. Arched windows are now very popular and, as a result, are usually very expensive.

Window sizes have also changed drastically over the past few years. For decades windows come in basic sizes. Today you can order virtually any size window you want for your house and the cost of special windows is not significantly higher than stock windows. However, if you can find stock merchandise that suits you, it might be wise to buy it because replacement, if needed, will likely be much easier.

You can buy windows that include the frames. If you buy the frames as well as the windows, your work is much easier when you are ready to install. If you bought frameless windows, you will need to construct frames.

Window framing stock varies, but usually 5-inch lumber is used for the interior framing as well as for the decorative or finished framing. You do not need to make any special cuts in the wood. Usually butt-jointed lumber works very well.

UNFRAMED WINDOWS

When you framed the walls of your house, you left rough window openings. Remember that these openings provided room for the window itself as well as about 2 inches of clearance. Before beginning you need now to become familiar with the various parts of a window assembly.

An explanation of terms

The window itself is made up of the *pane*, the glass that fits inside the wooden or metal or vinyl frame. The inner partition that crosses the window parallel with the ground is called the *horizontal bar*. There are usually two of these in an ordinary double window.

The vertical strips of wood, metal, or vinyl that run the length of the window are called the *muntin*. The thicker outside borders are called the *top rail*, the *upper stiles* (the borders on the sides of the window assembly), and the *lower meeting rail*. These are all included in the upper half of the double window.

The lower half of the double window is made up of the *upper meeting rail* (the point where the two window halves overlap), the *lower stiles* (one on each side), and the *bottom rail*. The entire assembly is referred to as the *sash* (FIG. 4-1).

The remainder of the window assembly includes the outside side casing, blind stops, side jambs, parting stops, side stops, and fasteners. These are all components of traditional windows; many modern windows do not include all of these.

The *side stop* is the small wood strip that is installed next to the sash to keep it from moving freely when the wind blows. The *parting stop* serves the same function on the back side of the sash. The *side jambs* serve the same function as the jambs on a door and fit inside the frame, actually enclosing the window sash unit.

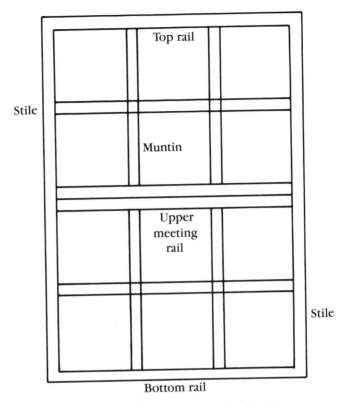

4-1 The components of a traditional window frame.

The *blind stops* are pieces of lumber that are installed between the side jambs and the outside side casings. These casings, which are installed both inside and outside, are commonly referred to as the framing.

Installation

When you are installing a wooden window, insert the upper sash into the allocated space to check the fit. Do not attempt to make a permanent installment at this point. If the window sash fits well, remove it and set it aside for the moment. If it binds, you can use a plane to shave small amounts of wood from the stiles. Try the sash often while planing. You do not want to remove any more than is necessary for a good fit.

If the sash is too loose, you will need to make the necessary adjustments in the side jambs. You can install a small shim between the wall framing and the jambs in order to make the fit together. This should take care of the problem for the lower sash as well as the upper sash.

If your assembly includes the parting stop (many modern window units have omitted this element of the window assembly), remove it so that you can set the window sash in its runway and slide it all the way to the top. While the sash is at the top, drive a small nail under the bottom rail and into the wide jamb to hold the window sash at the top.

4-2   These windows are fitted perfectly and the meeting rails are exactly where they should be. Note how carefully the frames and window units are fitted.

Set the lower sash into place and check the fit. If it is too tight, again use the plane to shave off just enough wood from the stiles to allow for a good fit. Set the lower sash into position and slip it into the runways. Let it move down until the bottom rail is resting on the sill. Note that where the meeting rails join they should be flush. If there is a small discrepancy, plane enough off the bottom rail that the meeting rails will fit together exactly (FIG. 4-2).

Replace the parting stop, if you have not done so already, and complete framing the window. It should now be ready for use.

Framing a window

After the window is installed, framing is the final step. The sole purposes of the framing are to conceal the necessary interior workmanship where the fitting occurs and to provide a more attractive and energy-efficient structure.

If you are planning to install paneling, wallpaper, or other wall covering inside, leave the framing work until the wall covering is completed. When you are ready to frame, work inside first and put up the vertical framing boards. These run from the exact top of the upper sash to the bottom of the lower sash. Usually 5-inch stock is used for this work.

You can use finish nails to install the vertical boards. Do not nail too close to the ends or edges. While finish nails are thin, they can still split wood if they are driven too near edges. Stay back at least an inch from any edge.

You will normally use two finish nails at the top, two at the bottom, and two in the middle. For aesthetic purposes, try to place the nails side by side.

When vertical boards are in place, prepare to install the horizontal boards. These should be long enough to reach from the outside edge of the first vertical board to a similar point on the other. You will notice that if you installed the vertical boards too high, your horizontal board will not fall low enough to cover the cracks formed by the interior installation work. In this case, you will need to remove the vertical boards and readjust them.

After the top horizontal board is in place, install the *sill*. This is a board usually 1 inch thick or thicker that runs across the window opening at the bottom. The purpose of the sill is to keep moisture from seeping through the cracks between installation timbers and boards and into the wall below.

The sill must fit well. Make sure that it completely covers the bottom of the window opening from front to back and on both sides. The sill should extend toward the outside far enough that any water that managed to find its way into the window assembly will drip off the back side of the sill and to the ground below. There should be a slight slant to the sill for this watershed purpose.

Inside, where the window sash fits against the sill, there will be a small crack—despite the most careful installation. When high winds blow rain directly toward the window, leakage into the house will occur unless you install an apron. The *apron* is a milled unit of lumber usually 2 to 3 inches wide. It fits into the window sash area and is snugged against each of the inside edges of the vertical boards.

Usually the apron is cut so that the inside edge fits between the vertical board edges, while the outside continues across the vertical boards and is aligned with the outside edges. Install this apron by driving finish nails down through the apron and into the wall framing below. Under the apron you can install the second horizontal board as you did the first. When you have finished, you should have a neat, tight, and geometrically perfect rectangle or square (FIG. 4-3).

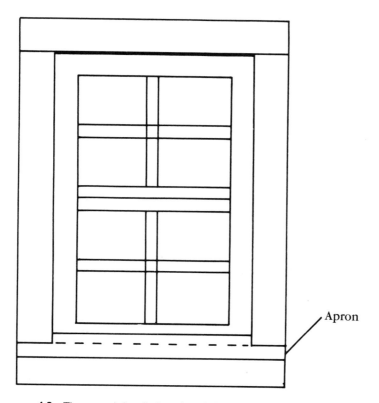

4-3 The apron is installed on the window to prevent leakage.

The outside framing of the window is more simply installed. While you do not install interior framing until the wall covering is done, you reverse the process on the exterior. You want the window framing to be completely installed before you begin to work on the wall covering. Whatever wall covering you are using will be fitted against the window framing, and efficient and neat installation requires that the framing be installed first.

Install the vertical side boards first, as you did inside. Then measure, cut, and install the top horizontal board. At the bottom of the window they will extend slightly past the wall surface, and you need to install the second horizontal board under the sill. Check to determine that the fit of the horizontal board against the bottom of the sill is tight. If it is not, insects and moisture can enter the wall easily.

Make sure that the blind stops are installed on the outside of the windows; otherwise the window sashes could fall out of the assembly.

FACTORY-ASSEMBLED WINDOWS

This is an easy job. When you buy used or old windows you run into complications with all of the unit members described above, as well as with springs and sash weights. New windows are manufactured without the springs and weights. When you buy them, complete with frames, all you need to do is set the units into the openings and fasten the frames to the wall framing.

When the window assemblies arrive, store them in a safe place until it is time to install them. Store them upright and safely protected from falling objects or errant timbers.

When you are ready to install the windows, measure the distance in the rough opening from side to side and from top to bottom. Then measure your window frames to be sure that they will fit into the openings. When you are satisfied that the fit is adequate, lift the assembly and set it into the opening. When the unit is in the opening, examine the window assembly for exact fit from a vertical and horizontal viewpoint. Center the assembly from front to back exactly, and when it is precisely where you want it, drive nails through the side jambs and into the wall framing. Take care that you do not hit the panes or stiles and mar them.

Space nails from top to bottom on both inside and outside of the window unit. When this is done, the installation is complete.

MAKING YOUR OWN WINDOWS

If you are enthusiastic about saving money and testing your own ingenuity and abilities, why not try to make a window? If the product is successful, you might want to make all of them. Making a window is a long and rather tedious task, but you can make one in half a day or less. As you become efficient, each succeeding window requires less and less time. When the work becomes tiring and exacting, remember that you are saving a considerable amount of money. If you can make two dual-pane windows in one day, you have saved yourself about $300.

There are several ways to make windows, but all require the same basic considerations. First, select the size window you want. Decide whether this size is for the entire assembly or the sash only. If you choose to make a double-sash window with dual panes for better insulation, you will need lumber stock 2 inches thick and 2 inches wide. You can use a 2-×-6 timber and trim it to your needs if you prefer.

Select the window size you want, but consider stock sizes of panes. You can cut glass to any size or shape you wish, but you will have less waste, less work, and spend less money if you use standard size panes. You will pay more for a 10-×-12-inch pane than for a 10-×-10-inch pane, and if you have to cut off 2 inches to make the first panes fit, you have wasted 2 inches of glass, its cost, and your work. You also run the risk of breaking a pane while you are cutting.

Assume that your windows will consist of two sashes, each with three horizontal panes and two vertical panes, each of which is 12 inches high and 10 inches wide. Your finished windows will be 30 inches wide (the space of three 10-inch panes), plus the width of the upper stiles (2 inches on each side), less the amount of space you recess the panes ($1/4$ inch on each side)—for a total of $33^{1}/_2$ inches of width. You can vary this space as you wish by using wider stile stock or larger panes.

Cut two stile pieces that are 27 inches long. This distance allows for two 12-inch panes (vertical length) plus 2 inches for the top rail and $1^{1}/_2$ inches for the upper meeting rail, less the $1/4$-inch recess in both the top rail and the upper meeting rail.

You can vary this length by allowing the top rail to run across the top of the panes and the upper meeting rail to run under the panes. If you choose this method, your vertical lengths will need to be 24 inches.

Your first step is to make a $1/4$-inch cut in the vertical stock so that the cut runs lengthwise one-third of the way from the edge of the lumber. Make a second cut two-thirds of the way across and along the length of the stock. When you cut the slots for the top rail and the upper meeting rail, as well as for the other side of the stile stock, you will need to have the slots meet at exactly the same point on all four corners of the window frame.

This will be a difficult task, unless you use the simple approach outlined here. Select a length of stock long enough to reach all the way around the window sash frame. Your length will include two 24-inch pieces plus two $33^{1}/_2$-inch lengths. Therefore, the total length of the stock should be slightly more than 115 inches. (The slight addition allows for sawing kerf or groove; you cannot saw without losing a fraction of an inch to sawdust.)

Cut the length of stock and lay it on edge, with the top edge representing the inside of the window sash. Measure over from one edge to a point where you want the first window panes to be located. The distance should be about one-third of the way across. Mark the point exactly at each end of the stock. Then use a chalk line and mark the line from one end to the other. Do the same at the two-thirds distance across the edge and mark the line with the chalk.

With the length of stock held firmly by a C clamp or other device, use the circular saw to saw a groove $1/4$ inch deep along the lengths of both chalk lines. These lines will represent the points where the window panes will be recessed into the stiles and top rail and upper meeting rail.

When you cut the stock into appropriate lengths for the sides, top, and bottom of the window, the grooves should be aligned perfectly—if you sawed along the chalked line accurately and if the stock was straight. You can now cut the stock for the stiles. Cut off two 24-inch lengths and lay these aside, grooved edges facing each other.

Cut the top rail and upper meeting rail next and lay these so that they face each other, with the two stile lengths forming a loose rectangle. Cut a 1-×-1-inch strip to serve as the horizontal bar. The strip may be larger if you wish to have more strength, but it should be the length of the top rail less the combined widths of the stiles. If the stiles are $1^{3}/_4$ inches wide, you will want to shorten the length of the horizontal bar by $3^{1}/_2$ inches.

Cut two strips of the same dimensions to serve as muntin strips. These should be as long as the stiles, less the combined widths of the top rail and upper meeting rail (FIG. 4-4).

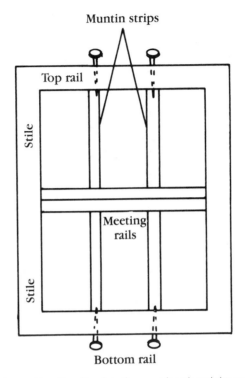

4-4   For muntin strips, cut lengths of lumber the exact length and size and space them one-third and two-thirds of the way across a window. Line up both top and bottom muntin strips and fasten these by using thin-shanked finish nails.

Lay the outside frame units so that they now fit together well and fasten them to form a solid rectangle. You can fasten the units by using brads or small, thin-shanked finish nails.

Lay the horizontal bar so that it is located exactly halfway between the top and bottom of the frame. Then lay the muntin strips at points exactly one-third and two-thirds of the distance across the window sash frame (FIG. 4-5). Be sure that the panes will fit once the horizontal bar and muntin are installed.

Insert the first pane into the upper left corner. Slide the edges of the pane into the groove cut earlier. Be sure the pane fits well into the groove in the top rail as well as into the stile.

Apply glue to the ends of the horizontal bar and to the points on the inside edges of the stiles where the bar should be installed. Let the glue begin to dry before positioning the bar. Hold the bar firmly in position until the glue sets well enough to hold the bar in its position. The bottom

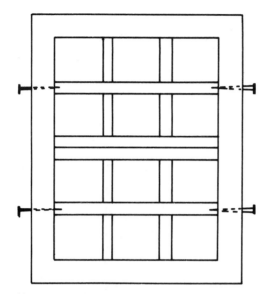

4-5 The horizontal bars are installed like the muntin strips. Cut the bars so that each section fits against the muntin strips or the stiles of the window.

of the pane already installed should rest firmly against the top edge of the horizontal bar.

Install the pane in the upper right corner and let it rest inside the grooves, with the bottom of the pane against the upper edge of the horizontal bar. You can also install the panes in the lower left and lower right corners.

Notch the horizontal bar and the muntin strip so that they will fit together where they meet, or cut the muntin strip and install one half of it at a time. Prepare the muntin strip at this point.

Mark the location of the first muntin strip and apply glue here and to the ends of the first muntin strip. When the glue has dried slightly, set the strip in position and hold it firmly until it has set. The strip should be against the pane of glass already set in the corner. Do the same with the remaining muntin stock. Check to see that both the muntin strips and the horizontal bar are firmly against the edges of the window panes.

You can now either cut a thin strip of wood to glue against the pane where it meets the horizontal bar and muntin strips, or you can use staples and caulk. You can also caulk where the pane edges fit into the grooves along the top rail and stiles and upper meeting rail. You can also use a combination of these methods. The caulk or putty serves to weatherproof the window. Use a putty knife to smooth the caulk to a professional-looking finish (FIG. 4-6).

Cut another groove in the stiles and rails for the purpose of adding a second window pane. Install the second panes as you did the first. When you have finished you have a double-pane window that serves both as window and as storm window. The dead air space between the panes will be good insulation.

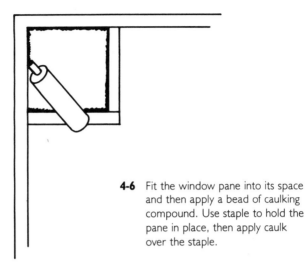

4-6 Fit the window pane into its space and then apply a bead of caulking compound. Use staple to hold the pane in place, then apply caulk over the staple.

NOTE: Clean the inside surface of each pane as you install it. Once the panes are installed, you cannot get to the inside areas to clean except by removing the panes. Once clean panes are installed, the inner surfaces remain clean because no dust or other particles can reach them.

While the methods of window construction given above seem very complicated and time-consuming, remember that as you continue to work you can build speed, and soon you can make a window in a matter of minutes.

DOORS

Approach door hanging in much the same way as window installation. The rough opening should be slightly larger than the door itself so that you have a small clearance area. Remember that, while windows come in a wide variety of sizes, door sizes tend to remain more traditional. The typical door is 35 inches wide and 6 feet, 6 inches high. The typical rough door opening is 40 inches.

The tools you will need include a hammer, level, and circular saw for the basic work. You will also need nails and a supply of lumber stock that can be used for door framing, jambs, and trim.

You can expect to pay from $200 up to $800 for exterior doors. You can shop around and perhaps find better bargains, but be sure you have found a real bargain before you make the final purchase. Ask what the door is made from. Check the thickness of it and the degree of finish on it. Ask whether the door comes with framing. Ask if there is a guarantee. A prehung door on steel mountings is a better door than one with no hardware or hanging materials.

Remember to examine doors for shipping damage as soon as they arrive. Store doors in a safe place where there is no danger of warping or damage. Do not store them where there is a possibility that people will walk on them. Keep the doors covered and protected at all times.

An explanation of terms

The basic terms in reference to door hanging include the following: header, top plate, top cap, head jamb, side jamb, wedge, and studs or cripple studs.

The *studs* are the 2 × 4s that are part of the wall frame. *Cripple studs* are short studs used below and above window and occasionally above doors. The *top plate* is the long 2 × 4 that runs across the wall framing, and the *top cap* is the 2 × 4 that covers the top plate (FIG. 4-7).

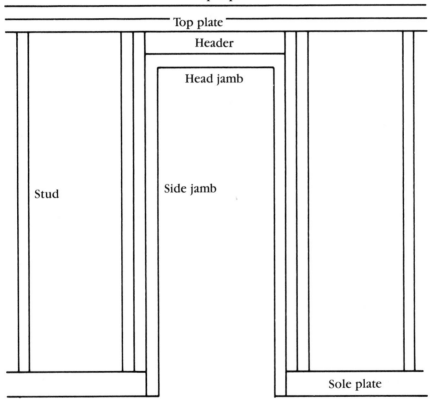

Top cap

Top plate

Header

Head jamb

Side jamb

Stud

Sole plate

4-7 A typical door frame unit. You can buy the precut jambs (with dado cuts) or make your own.

Headers are units of 2-×-6 stock fastened together with strips of plywood sandwiched between them. *Head jambs* and *side jambs* are the boards that form the interior surfaces of the doorway. *Wedges* are the shims or small shapes of wood that are used to secure the proper distances between studding and jambs.

Installation

Start your work by measuring the width and height of the rough door opening to be sure that the dimensions will accommodate the door itself. Then select the inside stock for the jambs. Side and head jambs are usually made of 3/4-inch-thick lumber that is 5 inches wide. These measurements are for interior doors. Exterior doors usually have jambs that are 1³/₈-inch stock. You can use other thicknesses without problems as long as your door fits.

Side jambs are usually cut with a dado near the top. The *dado* is a square-cut groove large enough for the head jamb end to slip into without difficulty (FIG. 4-8).

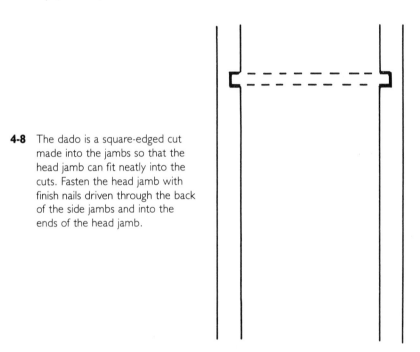

4-8 The dado is a square-edged cut made into the jambs so that the head jamb can fit neatly into the cuts. Fasten the head jamb with finish nails driven through the back of the side jambs and into the ends of the head jamb.

When you install the side jambs, use your level to assure yourself that the jambs are perfectly vertical. If the jambs are not true or perfectly vertical, the door will be either tight or loose at the top or bottom. If the jambs are installed correctly, the door will fit well along the entire length of the jambs.

If a jamb is out of vertical reading or out of plumb, use a wedge behind the jamb to adjust the jamb in the proper direction. To install the wedge, first measure the distance the jamb is out of plumb and select a wedge of appropriate thickness. Take the jamb down and set it aside while you tack the wedge to the door framing studs. When the wedge is installed, try the jamb again. Keep working until the jamb is plumb. Do the same with the other side of the doorway. When both jambs are plumb, install them by driving a small finish nail into the jamb and framing behind it. When each jamb is secured temporarily, recheck for plumb and then measure to reassure yourself that the door space is still exact.

When all distances are correct and plumb is maintained, complete the installation of the jambs by driving in finish nails up and down the length of the jambs. You are now ready to install the head jamb.

The side jambs with the dado cuts at the appropriate heights are already installed. Now slip the head jamb in place by sliding it inside the dado cuts and pushing it back until the edges of the head jamb are flush with the edges of the side jambs.

Some builders like to install the head jamb while the side jambs are still uninstalled so they can drive a finish nail through the back of the side jamb and into the end of the head jamb as it is firmly set inside the dado cut. Then the entire assembly is set into the door opening and installed. Secure the head jamb further by driving finish nails up through the bottom of the head jamb and into the bottom of the header, which is installed above the doorway.

The bottom ends of the side jambs are typically cut slightly short to allow for the installation of thresholds and carpet inside the doorway.

When the basic framing is completed, prepare to hang the door. First cut out the hinge inset. This is a small indentation in the edge of the door to allow the hinge to fit flush with the surface of the door edge. If you did not have this inset, the hinge would extend higher than the edge of the door and you could not get a proper fit.

To cut the inset, first place the hinge where you plan to install it and draw a pencil line around the outside edges of the hinge. Try to hold the hinge securely and check to see that it is placed squarely on the door edge. When the mark is made, you will use a chisel and hammer to cut out a section of wood the thickness of the hinge.

Start by holding the chisel on the pencil mark. Tap the hammer gently until the point or blade of the chisel is driven into the wood shallowly. Move the blade of the chisel over slightly and tap again. Keep repeating this process until the entire outline of the hinge has been covered.

Now tilt the chisel so that the blade is in the tiny cuts you have made and the top of the chisel is angled away from the hinge area. Tap the chisel with a hammer until the wood inside the outline begins to chip out. When you have completed a very shallow cut throughout the entire hinge area, smooth the surface of the cut and then lay the hinge inside the cut. If the hinge is not flush with the edge of the door, remove the hinge and cut again with the chisel. Fit the hinge frequently. Stop soon as the cut is deep enough for the hinge to fit flush with the door edge. Be careful not to cut the indentation too deep. If you do, the door will be marred permanently (FIG. 4-9).

The best location of the hinges is a subject of personal preference. Some people prefer three hinges, particularly for heavier doors, with one hinge in the middle and the other two about a foot from the top or bottom of the door edge.

4-9 Use a hammer and chisel to cut out a notch for the hinge to fit.

When the hinge insets are finished, set the hinge in the inset and insert the screws in the holes. You might find that the screws can be set better if you drive a finish nail into the space first and then pull it out. The finish nail opens a small hole in the wood and the screw can be driven more easily into the wood. You might also wish to drive a nail into the wood only to a depth of 1/2 to 1 inch, then let the screw force its way through the remainder of the wood.

Hanging the door

When the hinges are set on the door, it is time to mount them on the door jamb or its equivalent. Measure down the door to determine the exact point where the top of the hinge reaches. Then measure and mark the corresponding point on the door frame. Remove the pin, separate the hinges, and mount the second half on the door as you did on the door edge. When all hinges are mounted, hang the door by lifting it until the hinges mesh and the pin can be again slipped into place.

When the door is hung, swing it gently to see that it moves on the hinges easily and that it fits into the door opening easily. If the door rubs at some point, use your plane to shave off part of the door edge. Plane gently and slowly and remove only as much as is needed. Try the door frequently, and as soon as it closes easily, stop planing. You are now ready to install the latch receptacle on the door framing.

Push the door to a closed position and note where the latch strikes the framing. Mark the spot carefully. Then hold the *striker plate*, which is the latch receptacle, so that the latch position is lined up with it. Use a pencil to mark around the outside of the striker plate and mark along the inside of the striker plate as well. This is the area where the latch will seat as the door is closed.

Fitting the door

Now open the door and use your hammer and chisel again to cut out the wood inside the striker plate area. Chisel out enough so that you can off-set the striker plate so it will remain flush with the surface of the side jamb. The portion of the wood inside the striker plate latch chamber must be chiseled deeper, so that when the latch is engaged into the striker plate it cannot slip out if the wind blows or if the house settles slightly. In many older homes doors will open and close at all hours of the day and night because the latch no longer engages the striker plate chamber effectively.

When the striker plate is seated properly, install the screws that come with the plate. Tighten the screws firmly but do not overtighten. By doing so, you can strip the wood and then the screws will not hold.

When you are fitting your door, you might want the tightest possible fit without having the door bind or stick. You can achieve this if you take the door off the hinges, stand it on the edge where the hinges are located, and use a plane to bevel the edges of the door slightly. By taking off a slight amount of wood on the corner of the door you permit it to open and close freely and still fit tightly when the door is closed.

Trim work

Once the door is properly fitted, you are ready to install the trim work. The first trim strip should be installed when the door is closed to its proper position and the door stops exactly where you want it to stop. On the side of the door opposite the direction the door opens, you can install a trim strip that will be nailed, flat side to the wall, to the side jamb. Push the trim or molding into position by edging it as close to the door as possible without causing a bind later when the door swings open or shut. Nail the trim strip to the side jamb by driving small finish nails into the center of the strip every foot from top to bottom.

Also install the second trim strip when the door is closed. This strip will be nailed to the side jamb where the striker plate is located.

When the door is properly positioned—closed with the latch caught or seated in the striker plate—hold the trim strip so that it barely misses contact with the door. Use small finish nails and install the strip as you did on the other side of the door. At this point when the door is closed there should be no gaps between door and jambs.

You will notice a space of about $1/2$ inch or more under the door. Do not lower the door to try to close the gap. You now have only the sub-flooring in the rooms connected by the door, and when the finished flooring is installed the gap will be taken up by the flooring or carpet. In fact, if extra-thick carpet is installed, you might have to remove the door from its hinges and saw a fraction of an inch off the bottom.

To shorten the door, lay it on a smooth surface that will not cause scratching. Maintain firm pressure on the door while you use a circular saw to cut off the excess length. Remember that a circular saw causes splintering on the top side of the material, so you might have to do considerable sanding to remove the splinters. A handsaw is much slower, but splinters much less. If your patience and endurance will permit, use a handsaw rather than the circular saw. A hacksaw blade is even slower, but the cut is also very smooth.

Installing locks

Doors typically come from the dealers without any locks. You must install them yourself or pay to have the job done.

The major disadvantages of traditional locks is that they can be opened or "picked" quite easily and they can be locked accidentally. When you buy a lock, do not waste money on anything but a deadbolt lock. You cannot open a deadbolt without a key. If you misplace your keys in the house, you cannot unlock the door even from the inside, unless you buy a deadbolt lock with a thumb latch on the inside. Deadbolt locks sell for about $50. Installation is not included in this price.

When you set about installing a lock, you will find a paper pattern or template. Once you have decided where the lock is to be installed, place this template over the edge of the door and wrap it around the door so that all key points can be clearly marked. The template markings will indicate to you where you should drill the holes in the door face as well as in the edge of the door. Mark the locations carefully. The lock is designed so that there is room for only very minor adjustments.

NOTE: Locks are traditionally set a few inches under the knob, but this custom is actually a fairly recent one. For years locks were installed where they would be most convenient for the users. If you prefer the lock to be chest-high, and if the position is not a great inconvenience to others in the house, locate the lock to suit your own needs or preference.

Drilling the hole in the face of the door can be a difficult task as well as an expensive one. Drills, like circular saws, tend to splinter the wood as the bit emerges. You can do a great deal of damage to a door face during the drilling, and the splintered area is not always covered by the lock cylinder.

One way to avoid splintering is as follows: start the bit into the wood and maintain gentle but firm pressure on it. When the tip of the bit emerges from the other side—and you may have to pause in the drilling to check frequently—stop the drill and remove the bit from the hole. Then go to the other side of the door and start the bit in the small hole that had just been made by the tip of the bit. Continue drilling back the opposite way until the two larger holes are connected in the center of the door. No splintering will occur if the work is done in this fashion.

In hardware stores you can buy a set of drilling circles that are handy, but slow, to use. These circles come in a variety of circumferences. They have teeth on the bottom edge and the top edge fits into a device that is connected to the drill. In the center of the circle is a small bit that penetrates the wood at least an inch before the circle does. Such a drill attachment is very handy and can be used to drill holes without damaging the door face.

When the hole is through the door, drill a connecting hole starting at the edge of the door—in the spot indicated on the template.

When you mount the lock you will find two long, slender bolts that extend through the cylinder and through the door and into the cylinder on the opposite side of the door. Each bolt starts from a different side of the door, making it impossible for an intruder to remove the bolts and gain access to your home. Thus the name "deadbolt." The bolt is connected to the cylinder mechanism as the metal extensions pass through the deadbolt assembly. When the key is turned, the mechanism on the inside of the door is engaged and the bolt is either shot or withdrawn. The entire assembly can be tightened firmly and the lock covers are forced into a tight position as the bolts are turned.

When you are finished with this stage of your work, all that remains is to install the striker plate.

Making a primitive window lock

Most windows come equipped with a lock that consists of a latch mounted on top of the meeting rail that engages an assembly on the bottom rail of the upper sash. Such locks are not fully efficient. Anyone seeking to break into the house can break the window and turn the latch and the window will open freely. The major safety factor here is that the noise of the breaking glass will alert family members.

You can make a slightly more effective primitive lock by drilling a small hole through the meeting rails of the window and installing a nail or large screw. Choose an unobtrusive location, such as the sides of the window where curtains will cover the nail or screw head.

When you drill the hole, select a bit that is only slightly larger than the shank of the nail. Do not drill the hole all the way through both meeting rails. Drill through the first rail and halfway through the second rail. When the hole is ready, you can insert the shank of a large nail into the hole. If the nail is so long that the head protrudes, use a hacksaw to cut off part of the nail until the head is flush against the rail surface and the shank of the nail extends well into the second rail. Repeat the process on the other side of the window.

If an intruder tries to gain access to your house via the window, he will have to break at least two window panes. Chances are good that he will not be able to see the nail heads and therefore will not know where the window lock is located.

In order to "unlock" the window all you need to do is withdraw the nail. After the window is opened you can insert the nail temporarily in the hole through the first rail.

For a more difficult "lock," use a large screw that must be inserted using a screwdriver. An intruder will not only have to locate the lock but he will also have to extract the screw before the window will open.

These locks are not burglar-proof; few locks are. The major function of such a lock is to cause sufficient delay or noise that the intruder will become either frightened or impatient and apprehensive and will abandon his efforts.

MAKING YOUR OWN DOORS

If you are building a rustic house and want an old-fashioned look, you can make your own doors easily. By doing so, you can save $300 or more.

You will need a chain saw and the correct type of wood. While oak has been the preferred wood for years, remember that of all the doors in the dealer's showroom very few of them will be solid oak. Doors are made from pine as well as from many hardwoods. Poplar makes a fine door.

When you locate the wood—whether you harvest a fallen tree or buy the wood—you can either saw the vertical boards with a chain saw or take the timber to the nearest sawmill and have it cut for you. Cut the log into the proper length, usually about 6 feet, 6 inches. You might want to add a few inches to allow for waste.

Lay the log on two built-up pillars to keep it off the ground while you work. Make a chalk line down the center of the log lengthwise. Use the tip of the chain saw to make a groove about 1 inch deep along the chalk line. Return to the beginning of the log and saw deeper until you have halved the log lengthwise. Make two chalk lines next, one along each edge of the log lengthwise and cut off these slabs by running the chain saw along the chalked line (FIG. 4-10).

Turn the log so that it stands on one of the newly cut edges. Measure over 3 inches, or whatever thickness you want your door, and mark the bark. Do the same at the other end. Then make a chalk line from mark to mark. Using the 1-inch groove as a guideline, saw through the log. Take great care to hold the chain saw vertically so that the log will not be thicker on one edge than it is on the other. When the cut is finished, you will have thick timber from which to make your door. Make other cuts until you have enough timber boards to equal 36 inches across and 6 feet, 6 inches in height.

At this point you can saw the logs so that they are uniform in length. You might want to plane the edges to get the smoothest possible cut. If you use a ripping chain on the saw you will have a cut that is as smooth and straight as anything you can buy in the lumber yard. A chipping chain or crosscut chain tends to make a much rougher cut.

You will need to have the edges of all timbers as straight as possible. Use a circular saw for the final edge cuts. When you have even cuts at all points, lay the boards in position and about 6 inches from the top, nail a board across them at right angles. Nail on another board about 6 inches from the bottom. Then mark, cut, and install a diagonal board so that you have the classical Z outline on the door (FIG. 4-11).

The door is now finished. If you prefer a smoother finish, plane all surfaces, then use a coarse sandpaper followed by a finer sandpaper. You

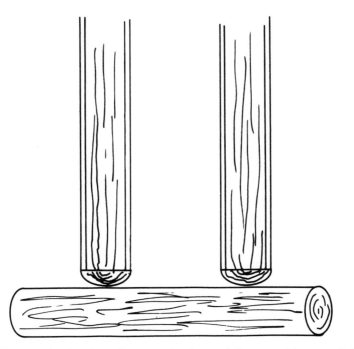

4-10   After you rip a log down the center, lay the two halves—cut side up—and rip the sides off. You are now ready to saw usable lumber.

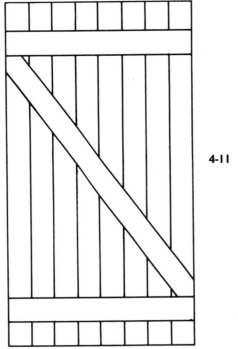

4-11   This type of door can be constructed easily and cheaply.

can also buy a sandpaper disk to attach to your electric drill. This will smooth your door easily and effortlessly, and will spare your fingers the abrasions.

For doors it is best to use boards that are 6 inches wide or nearly so. The intended look of such a door is that of the old log cabins. You want it to be slightly rough and irregular.

You will need to air dry the door for several weeks, so you might want to construct it before you do the rest of the finish carpentry on the house. After the door has been sanded and dried, you can stain it.

Door construction using rod

It is possible to construct the door another way by using threaded rod. This rod can be bought at most hardware stores and is the rough equivalent of a yard-long bolt, except that it has no head. You will need three of these rods, one for the top, bottom, and center of the door.

First drill a hole large enough for the rod. The hole must go through each door timber from edge to edge and in the center of the edge so that there will be equal strength all around the door. Stand the boards on edge and drill the holes. When the first hole is drilled, position the drilled board atop the edge of the second board. Run the drill for a second or two through the first and into the second board so the two holes will align exactly. Use the same technique with all subsequent boards.

The holes should be slightly smaller than the rod, if you can find the appropriate bit, so that when you drive the rod into the holes the fit will be very snug. Even when the door air dries the fit will be a good one (FIG. 4-12).

Insert the rod through all the aligned boards. Do not drive the rod by hitting it directly with a hammer. Lay a block of wood on the end of the rod and hit the wood with the hammer. The force of the hammer will ruin the threads of the rod.

When the rod is seated, tap it an inch or so deeper so that one end extends through the final board. Use the circular drill attachment to drill a hole large enough so that you can countersink the nut and washer for the rod. When the hole is ready, tap the other end of the rod until it is in the right position and fasten the washer and nut to the end. Do the same at the other end of the rod and start the second washer and nut.

For best results the rod should be slightly shorter than the width of the door. If the door is 36 inches wide, the rod should be 35 inches long. There will then be $1/2$ inch of leeway at each end of the rod—enough space for the washer and nut.

When all three rods are inserted and completed with washers and nuts, tighten the nuts until they pull the wood tightly together, but do not apply so much pressure that you crush the wood fibers. When this is done, set the door aside to air dry. Put it where it receives the full benefit of air and sunlight, but is protected from the rain. Every week tighten the nuts slightly, so that as the wood shrinks as it dries, you maintain constant pressure on it. When the door is dry enough so that the nuts will not turn readily, it is ready to hang.

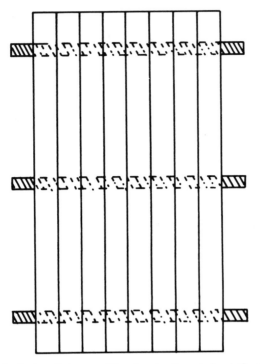

4-12 Drill the holes in door timbers and then insert a threaded rod. Countersink a nut on each end and then cover the hole with plastic wood.

Such a door will lend a charming effect if your house is a rustic style. It will also make a fine door for workshops and similar outbuildings.

If you want the rustic effect but do not want to spend so much time on the project, use lumberyard boards that are either 5½ inches wide or wider and create your door from these. For best results the doors should be about 2 inches thick.

STORM WINDOWS AND DOORS

Perhaps you want to add storm doors and storm windows to weather-proof your house as much as possible. These materials come in a wide range of prices, some windows as low as $25 and as high as $75. You will have the option of buying the regular windows or the highly publicized triple-track windows that cost considerably more.

Windows

There are advantages to the regular windows, and many people prefer them to the most expensive triple-tracked counterparts. Ask a salesman to show you the advantages and disadvantages of both, then buy the one of your choice rather than the one that is most highly advertised.

Before buying, measure your windows to be sure of getting a perfect fit. Measure from the inside edge of the window framing, both from side to side and from top to bottom. Do not measure only the width and height of the window. The storm window will be attached to the framing and must be large enough to cover the entire area inside the framing.

NOTE: When installing storm windows, place a ladder close enough to the window so that you can work comfortably without endangering yourself. If you must work at considerable height, have someone on the ground to hold the ladder to stabilize it.

Many storm windows snap out of the frames. If yours will, you can install only the frames. The load will be much easier to manage, but you must be careful not to warp or twist the window frame even slightly. If you do, the windows will not fit properly. One way of handling the problem is to install the frame partially by using a screw in each of the four corners. The screws will hold the weight of the frame and windows as you install them. If the windows fit properly, install the remainder of the screws.

You can make your work easier by using either a drill with a very small bit or a hammer and nail with a thin shank. Use the bit or the nail to make a small hole so that the screws will start easily. Then tighten them fully after you are able to use both hands.

After the window is installed, use a caulk gun and fill the cracks between the frames of the storm window and the regular window. Apply a good thick bead along the crack, then use a putty knife to spread the compound. Feather it so that when the window frame is painted the paint will conceal the putty.

If you cannot find storm windows that are long enough to fit your traditional windows, you can make minor adjustments. If the storm window is an inch or so too short, cut and install a board that lies atop the window sill but does not extend far enough toward the inside to interfere with the opening and closing of the window. The thickness of the board will allow the storm window frame to cover the window area and you can insert the screws into the newly installed board rather than the old sill. The new board will not be visible from the outside, and it is not noticeable on the inside of the house.

Doors

Install storm doors the same way you installed storm windows. Measure from the inside edge of the door framing, from side to side and from top to bottom. The frame of the storm door will be held in place by a series of small screws that extend from top to bottom on both sides and that also extend across the top of the door.

As you did before, hold the door in place until you can seat three or four screws: one at each top corner and one or two at the bottom framing. When the screws are fully seated, try the door to see that it opens and closes properly and that the inside glass and screen can move as they should.

Again, use a drill and small bit or hammer and nail to make starting the screws much easier. When installation is complete, use the caulk gun to seal the cracks.

If the door is too short, you can "lengthen" it as you would a storm window, except in this case you must install the added strip at the top rather than at the bottom. Cut a board the same length and width as that of the head jamb. Before you can install it you will need to remove the molding strip across the head jamb and those on the side jambs. The molding on the side jambs will have to be shortened by the thickness of the new head jamb board. When the jamb moldings are removed, install the new board by driving six nails, two at each end and two in the middle, up into the old head jamb. Then reinstall the molding on the head jamb and cut off the thickness of the new jamb from the side molding and renail the molding (FIG. 4-13). Lower the storm door frame so that the top can be attached to the new head jamb board. Proceed as normal along the rest of the framing.

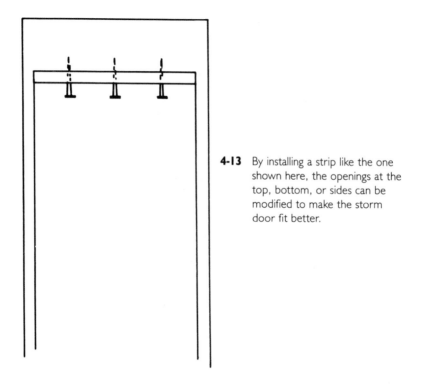

4-13 By installing a strip like the one shown here, the openings at the top, bottom, or sides can be modified to make the storm door fit better.

The final step is the installation of the closer. Simply screw the mounts in place, according to instructions, and attach the closer to the mounts. You have an adjusting set on the bottom of the closer of most doors. You can loosen or tighten the set so that the door will close to your preference.

You will also need to install the latch for the door lock. The latch is made up of an extending piece of metal and a flat expanse of metal that has two angled slots in it. These slots allow you to slide the latch up and down to make it fit the lock latch. You can also tighten the closure of the door. Adjust the latch base slightly by loosening the screws and sliding the latch assembly downward. As you do so, the latch catch is moved farther from the latch itself so that the door closes tighter when the latch is caught.

SIMPLE SHUTTERS

You can make simple shutters for your windows. All it takes is a few minutes and a small amount of lumber and effort. These shutters are not functional; they are purely decorative. Storm windows have all but eliminated

4-14 Shutters, although no longer functional items, can add to the appearance of the house.

the need for functional shutters, and many houses no longer have room for shutters. If your house does have room, a simple shutter will break the empty space on one side of the house and provide variety in appearance (FIG. 4-14).

Materials

The materials these shutters require is 1-×-4 lumber, a saw, hammer, nails, and a square.

Begin by measuring the window from top to bottom. Add 10 inches (5 for the top and 5 for the bottom)—if the window does not have any projections such as unusual window sills or frames that would render the extra length impractical. If there are window interruptions, shorten the shutter length to that of the actual window area, plus an inch or two at the top. If the window is 56 inches in height, you will need three boards of that length for each side—if both sides permit your using the space. If not, make and install the shutters on only one side of the window. If you have double windows, add the shutters on each side of the window.

Making the shutters

Lay the three boards on a good work surface. If one surface of the boards is better than the other, turn the better surface up. Spread the boards so that you have 2 inches between them. If space is a problem, reduce this to 1 inch. Mark a point 10 or 12 inches from the top and from the bottom of the boards. Measure the distance across the three boards and cut a short board that will reach across the three boards. You can let the cross member extend an inch or two on each side if you like (FIG. 4-15).

Install the cross member either under or on top of the boards. Try it both ways to see how it most appeals to you. When you have the assembly at the best location, use small finish nails to drive through the cross

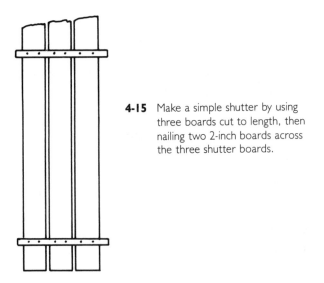

4-15 Make a simple shutter by using three boards cut to length, then nailing two 2-inch boards across the three shutter boards.

piece and into the longer boards. Attach one near the top and one near the bottom. One-fourth of the way from each end works well.

When the shutter is finished, you can paint or stain it. In a few minutes you can make such shutters for the entire front or side of the house.

You can build functional shutters if you have room for them. You will need more lumber as well as installation room. Build them the same way as just described, but measure from the end of the window to the center and make the section of each shutter the size needed to join in a loose fit in the center of the window. Add a latch assembly halfway down and you can close the shutters tightly in threatening weather.

Use hinges such as those found on a large cabinet door, so the shutters can swing closed. Install a vertical board beside each window and attach the shutter hinges to the vertical board.

Installation

To install either type of shutters, use long screws or finish nails. If the shutter material is very heavy, use larger nails with heads. If you have wood siding or vinyl, you can drive the nails through the shutter vertical board and into the framing under the wall covering and the shutters will hold well. If the siding is brick, you can buy masonry nails with which to install the vertical boards, one on each side of the window. You can also drill a small hole and use screws to install the vertical board.

Chapter **5**

Finishing ceilings

As with most of the work that goes into finishing a house, you can spend a great deal of money and devote seemingly endless hours of effort to a project, or you can simplify the goals and spend far less money and time. Ceilings are no exception.

There are various materials you can use to finish ceilings: beaded board, tongue-and-groove boarding, ceiling tile, Sheetrock, wallpaper, to name the most common. Sheetrock, because of its weight and size, is among the hardest surfaces to install alone. Ceiling tiles are among the easiest. The latter are also among the most expensive if you buy high-quality tiles at regular prices.

In this section you will receive several suggestions on how to accomplish a finished ceiling look with a minimum of trouble and with reasonable expense.

BOARD CEILINGS

One of the most beautiful ceilings is that made from knotty pine or similar lumber. For many years this type of ceiling was popular; it fell out of favor in part because dust and dirt filtered through the cracks between the boards and into the room below, where it littered food, furniture, and clothing.

There are ways to avoid such problems. You can apply plastic or sheathing paper to the ceiling joists before you begin covering the ceiling. You can also use tongue-and-groove lumber which, when tightly fitted, forms a seal that prevents virtually all dust and dirt from filtering through.

Choosing a pattern

To install traditional boards, first measure the room and buy your lumber in accordance with the dimensions of the room. You can either cover the entire board course with one timber; or go with an alternating whole board and two halves, followed by another whole board; or use variations of patterns involving entire lengths, followed by pieces course; or use all pieces courses. For the most attractive look, establish a pattern and stick to it all the way across the ceiling. Do not do a random fitting (FIG. 5-1).

If your room is small, you might want to use whole boards. You can buy an adequate supply of 12-foot boards and nail them in place side by side until the ceiling is covered. A good look is created by using one 12-foot board followed by two 6-foot boards, followed by another 12-foot board, followed by a 6-foot board in the center and two 3-foot boards at either end.

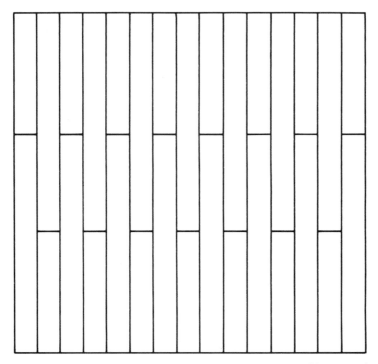

5-1 Save on lumber and retain a symmetrical look to a floor or wall by staggering the lengths of boards. In this way both long and short lengths can be used.

Installation

As you begin work, have the joists or rafters already in place and ready to be covered. If you are using tongue-and-groove boards, you probably want to start with the groove side of the first board against the wall and the tongue sticking out toward the center of the room. This way you will have a nailing surface.

With the first board in place, start your first nail about 1 foot from the end of the board. It is very helpful if you have someone to hold the other end in place while you drive the first nail. If you are working alone, you might need to drive the first nail in the center of the board so there will be less board to suspend on one end.

Use cut nails or square or rectangular nails that are made especially for use with tongue-and-groove lumber. Set the nail so that the broad side is parallel to the floor and ceiling and the point is in the angle formed by the tongue and the cut edge of the board. Before you drive the nail, check again to see that the board is pushed firmly against the wall and that it is even in both corners. It is very difficult to remove tongue-and-groove lumber without damage.

When you are ready, drive the nail until the head is even with the imaginary line formed by the third part of the triangle, which is formed by the tongue and edge of the board. Do not try to sink the nail the rest of the way; you will crush the tongue and damage the board so that the groove of the next board will not fit over the tongue well (FIG. 5-2).

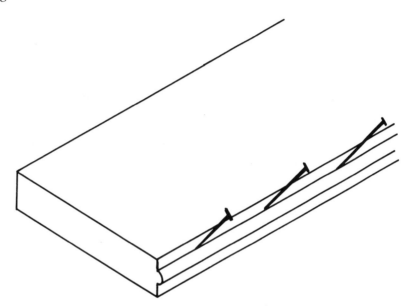

5-2 To have concealed nail heads, drive the nails at an angle just above the upper edge of the tongue. Use a punch to seat the nails the remainder of the way.

When you have driven several nails in this fashion, use a punch to sink the nails the rest of the way. The punch has a point that can be placed on the nail head so you can hit the other end of the punch and drive the nail all the way in. You can also lay the thick part of the punch—the part that is usually scored—across the head of the nail and then hit the shank of the punch with the hammer. This latter method works best if you can manage it.

Complete nailing the first board. Then hold the second board so that the groove faces the tongue, and starting at one end, slip the groove over the tongue. Hold a length of board that is a foot or so long against the tongue of the second board and tap gently until all of the board has been seated properly. The groove should at this point completely cover the tongue so that no part of it can be seen. There should be no cracks between the first two boards. Nail in the second board. Always use the punch for the final seating of the nail head.

Proceed in this way across the entire ceiling until you come to light fixtures or wires where fixtures will be attached later. When you encounter lighting wires or boxes, make certain that the power to the wires is off. Hold the next board so that it partially covers the box. Use a pencil to mark a semicircle, or whatever the shape of the box, on the board. Trace the outline of the box as accurately as possible. Then remove the board and cut out the traced shape. Install the board as you did the others.

At the next board trace the remainder of the box if the board is wide enough. If it isn't wide enough, cut the board in half so that the two halves will meet at the box. Then hold up one half at a time and trace the outline of the box on each half. Cut out for the outline and install the half board. Do the same on the other side of the box. Continue in this fashion until the entire box is boarded around, then proceed to the other wall.

When you are installing square-edged boards, all you need to do is staple up plastic or sheathing paper before you begin work. Then install the boards by shoving the first one firmly against the wall and nailing it in place. Position the second board and all subsequent boards firmly against the previous one and nail them in place. You will perhaps want to start your nails while the board is on the floor and turn it so that the nail heads are down. Then you can hold the board with one hand and hammer with the other.

Treat the electrical fixture box in the ceiling the same as you did the tongue-and-groove boards. Regardless of what type of boards you are using, when you come to the end of the ceiling, you might need to cut a board to fit it into the final space. Do so by measuring the space between the last board installed and the wall. Mark and rip the board and then install it. If the final board is tongue-and-groove, you might need to make it slightly smaller so that you will be able to work the groove over the tongue before the final nailing is done. If you find that you cannot wedge the groove over the tongue, trim the tongue off the final piece installed and work the partial board into the space left. The absence of the final tongue will not be a serious problem.

You can also install the boards in a diagonal fashion or mix the arrangement to suit your own tastes, just as you would with wall coverings. Create your own patterns for interesting variety.

DRYWALL CEILINGS

This is almost always a two-person job. A panel of drywall or Sheetrock is 4 feet wide and 8 feet long and very heavy for one person to hold in place while nailing. Unlike boards, which can be held by one nail, you will need to put several nails into the drywall or the weight of the panel will pull the board over the nail heads.

Drywall is fairly easily damaged, so you don't want to risk dropping the panel. It is also heavy enough that you can hurt your back by lifting and holding the weight for a prolonged period of time. Also be aware of the danger of the partially nailed panel falling upon you as you work. Some methods of installing drywall singlehandedly will be provided later.

Before you begin to nail up drywall, first check to see that the ceiling is ready. Measure to see how many panels will be needed to reach across the room and in which direction the panels should be installed. If your room is 20 feet wide, you can put up two panels lengthwise and one panel crosswise and that part of the room will be covered. This is a better method than putting up a full panel then cutting a panel to get the remaining half.

You will not always be able to deal with full panels. If the room is a length or width divisible by four, you can work out the arrangement easily. For other dimensions, you will be forced to cut a panel. Remember that you can cut the panels with a handsaw or with a sharp knife that has a sturdy blade.

Installing drywall with a helper

If you have a helper, arrange the work platform so that you can both reach the ceiling easily. Each of you should grasp one end of the panel and lift it to chest height. Then arrange your hands so that you can comfortably push the panel up until it reaches the ceiling. Push the panel firmly against the corner of the ceiling and wall. While one person holds the panel in place, the other can begin to drive the nails through the panel and into the ceiling rafters or joists. You will need to work rapidly because of the weight on the other person, but do not be in such haste that you damage the Sheetrock face by hitting the nails too hard.

When the first panel is installed, move the work platform so that you can install the second panel. Lift it and nail it as you did before.

Measure to the center of the room where the light fixture or box will be located, if there is one in the room. Try to work out the installation pattern so that the fixture box will be at the edge of one of the panels. If you cannot do so, smear chalk on the edge of the fixture box and lift the drywall panel into position and press it firmly against the fixture box. When you take the panel down, the outline of the box should be clearly visible on the back of the Sheetrock panel. You can now cut out this section. Reposition the panel over the box and nail it in place as you did the others.

When installation is complete, you will need to apply the compound and tape. When this is done, sand the nail coverings and prepare the drywall for painting. (See Chapter 2.)

Installing drywall singlehandedly

This is not a recommended process, but with patience and hard work it can be accomplished. Your job is to somehow lift the drywall and hold it in place while you nail it to the ceiling rafters or joists. The easiest way to accomplish this is to construct a helper.

The "helper" consists of several short lengths of 2 × 4 and one longer 2 × 4. Select a 2 × 4 that is 2 to 3 feet long and stand it on edge. Abut a 2 × 4 that is 8 feet long and also standing on edge into the center of the first length. Nail up through the bottom of the first 2 × 4 into the end of the longer one.

Measure the combined distance from the base length to the end of the longer timber. Then measure the distance from the floor to the ceiling. You will need to add another length of 2 × 4, this one 3 feet long or slightly longer, to the free end of the longer timber. You might need to shorten the longer timber slightly in order for the entire assembly to stand against the wall with the thickness of one panel of drywall on top of it (FIG. 5-3). In other words, if your ceiling is 8 feet high, the assembly you have put together must be 7 feet, 11½ inches tall, with just enough room for the drywall to be placed upon it. If you wish you can also add a support piece to the front and back of the first 2 × 4 you used. This device will stand alone and will support the weight of the drywall.

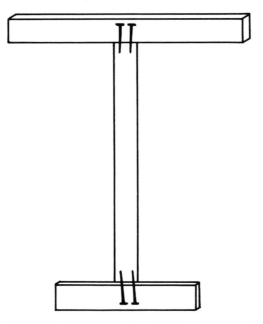

5-3   This device will help when installing drywall singlehandedly.

When you are ready to work, stand the holder against the wall or wherever you are intending to work. Lift a panel of drywall and slide the end of it over the 2 × 4 that is parallel to the ceiling. You can position the stand or holder so that half of the drywall is resting on it. Now all you have to do is lift one end of the panel a fraction of an inch and start the nails. When you have worked your way around the entire panel, it will be firmly nailed and you can move the holder to the next work area.

Do not shove the drywall roughly across the holder and do not leave the drywall suspended on the holder while you leave the work area. This device is merely a helper and can easily be dislodged or tipped over while you are working. As such it serves as an emergency measure only and works well enough to get you through the roughest part of the job.

CEILING TILE

Ceiling tile can be bought for as little as $.30 per square foot or more than $1 per square foot. Consider that if your room measures 12 × 14 feet, you have 168 square feet to cover. This means that you might pay $168 or more for the ceiling tile for that one room. This is not very expensive as building materials go, but it is generally wise to stop and examine the economy of a project before beginning it.

Installation

Installing ceiling tile is not a difficult job. You will need only a knife and a staple gun with plenty of staples. The tiles are manufactured with a lip on one end and an indentation on the other. These lips and indentations serve almost like tongue-and-groove lumber. When you install the first tile and staple it securely, the second will have a lip that fits into the indentation of the first. You staple the second tile and fit the third into the lip of the second. In this fashion you work your way across the room.

When you are ready to start the second row of tiles, they must face in the same direction as the first row, and you will need to cut one tile in half. This half tile at the start of the row will cause the joint line to be staggered, just as the masonry bond line. By staggering the tiles, you will not have the weakness in support that would occur if all the tiles ended at the same point (FIG. 5-4). Because you staggered the tiles at the beginning of the row, you will need another half tile at the end of the row. You can use the other half of the tile you used to start the row. On the third row you will start with whole tiles, and on the fourth you will need to cut another half tile and stagger the installation.

When using the staple gun, hold it firmly against the surface to be stapled and try not to move it as you release the staples. Try to seat the staples firmly and far enough back from the surface that the entire staple is embedded in the tile material.

As you work from row to row of tiles, look back regularly to see that you are keeping the full tiles and half tiles aligned perfectly. Even a slight deviation can be noticeable later.

When you come to the far wall, if you see that you are going to need a partial tile row, save out any tiles that have been slightly chipped or marred during the work period. Use these tiles for the partial rows by cutting off the edge where the damage occurred.

Tiling around light fixtures or connection boxes is easy with ceiling tiles. You can press the top side of the tile against the metal framing of the box and a clear imprint will be made. Use a knife to cut out for the box, then tile around it without difficulty. Later, when the fixture is remounted, it will be impossible to tell where the fitting cut was made.

Before you begin any ceiling tile installation, check to see that the joists or rafters are spaced properly to allow you to fasten the tiles acceptably. You will probably have to install furring strips for adequate installation. You might find that the tiles need to be installed at right angles to the direction of the rafters or joists. When you install furring strips, use 1/2-

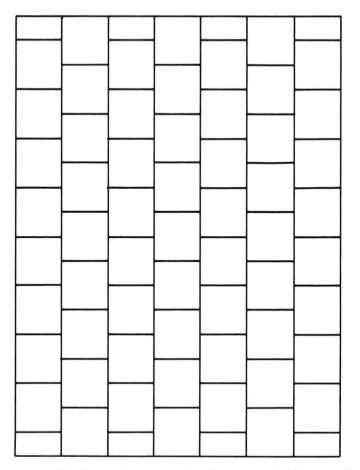

5-4   Stagger the ceiling tile units for a neater look and more secure holding power. A partial piece is needed on the ends of alternate courses.

inch lumber 2 inches wide. If you don't have ready access to such lumber, rip a 4-inch board down the center and use the two strips. One major consideration is that you must keep the furring strips the same thickness. If you do not, you will have an uneven ceiling that is very unsightly, even if the discrepancy is only a fraction of an inch (FIG. 5-5).

Before you choose a ceiling tile, shop around and check the various samples that are available. You can buy tiles in many patterns and thicknesses. Also consider acoustical tiles.

SUSPENDED CEILINGS

At times an existing ceiling is so badly defaced and uneven that tiling on top of it would be impractical. Or perhaps you want to lower a ceiling to save on power bills.

Today, the old-fashioned high ceilings are totally impractical, except for aesthetic purposes. Consider that a room with a 12-foot ceiling con-

5-5 Install furring strips as shown above if the ceiling is not even or level.

tains 50 percent more square footage than a room with an 8-foot ceiling. This extra footage is useless under average circumstances. You will be paying a heating bill that is 50 percent higher than necessary and not realize any extra comfort for the extra money. You will also spend 50 percent more money on paneling, wallpaper, plaster, or any other type of wall covering. Not only is the expense greater, the difficulty of working at such heights is also a negative factor. Even a 10-foot ceiling results in a 25 percent increase in heating costs and construction costs. Lowering the ceiling is a wise economical move.

There are two basic methods of lowering ceilings. The quick and easy way is to buy and install wall angle materials, then install the main and cross trees. The wall angles are metal strips that can be nailed to the existing wall, and each strip has a right angle piece that juts out into the room an inch or two. The main and cross tees are similar metal strips that dissect the room into 2-foot squares or 2-×-4-foot rectangles. When the main and cross tees are installed, all you need to do is lower the tiles into the receptacles formed by the tees and the wall angles. You do not fasten the tiles in any way, and if at a later time you choose to remove the tile to change patterns, all you have to do is lift out the tiles.

You can buy the gridwork needed in either steel or aluminum. Steel has the advantage of greater strength, but the aluminum is much lighter and there is seldom any great stress placed on the gridwork under normal circumstances.

The hard way to lower a ceiling is to nail 2 × 4s around the wall to form a ceiling line and then to install 2 × 4s across the wall so that they abut the ceiling line timbers. This is equivalent to installing new ceiling joists, except that you use 2 × 4s rather than 2 × 6s or 2 × 8s (FIG. 5-6).

Space the joists so that they conform to the width of the ceiling tiles. You might choose to add cross pieces between the joists for a bridging effect and to provide a space to staple the ends of the tiles.

Either way works well. The major difference is that the second method allows no way to lift out the tiles and replace them without damaging them.

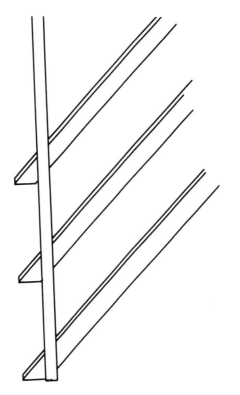

5-6 A method of lowering a ceiling. Use the bottom portions of the boundary timbers as molding.

CREATING DECORATIVE CEILINGS

You can achieve a unique ceiling look by applying a caulk-like compound with a special roller. This is easy to do and the results are usually very pleasing.

If the ceiling already has Sheetrock or boards on it, you can apply the compound easily. Dip the roller into a roller pan filled with the decorative compound. Hold the roller firmly and push it with modest pressure into the compound. You can tell easily when the roller is "loaded" fully. It becomes heavy and soggy. Lift the roller to the ceiling and start at a point about 18 inches from a wall. Push the roller against the ceiling with very faint pressure. As you begin to move the roller, apply gradual pressure until the compound starts to spread. The roller itself has erratic grooves in it and these will "wobble" through the compound and create unusual designs.

Apply several rollers filled with the compound until you have reached the desired thickness and design. Apply the compound thick enough that the Sheetrock or other ceiling material cannot be seen through it.

Work the roller back to the wall as you proceed. You will not be able to spread the compound all the way to the wall. You must take down the ceiling molding and store it in a safe place until the compounding is complete. Also take down any light fixtures or other items that may interfere with your spreading the compound.

When you have finished the ceiling will be spackled with an attractive design. The only problem is that the ceiling will be a glistening white. You cannot paint the hardened compound in a traditional way. You can use a spray paint, but you must be careful not to spread the paint to the wall. Ask your dealer if he has a dry paint mixture of acrylics that can be mixed with the compound.

NOTE: If you have a leak, the water seeping through the compound will cause it to disintegrate and discolor. The dried compound will also chip and flake if hit or scratched. Such ceiling coverings will not tolerate rough treatment.

PAINTING CEILINGS

Paint provides one of the very best ceiling decorations possible, assuming that the ceiling is in good physical shape in other respects.

Gone are the days when the only paints available were the oil-base products that smelled up the entire house and required days to dry. Modern technology has produced paints that have no offensive odors, dry in minutes, can be cleaned up with a damp cloth, and will not drip or run while drying. Perhaps best of all, these paints come in an almost infinite array of colors and shades so you are sure to find the color you are looking for.

When you are preparing to paint, take a few extra minutes to remove the ceiling molding; cover the floor with a drop cloth; and move any valuable furniture, paintings, or art work out of harm's way. Water-based paints can be cleaned easily and quickly with a damp cloth while the paint is still wet. However, if a drop of paint falls unseen on a valuable item of furniture and remains there until it is thoroughly dried, it is very difficult to remove without doing damage to the surface of the furniture.

Stir the paint well and pour it from one clean bucket back into the original can repeatedly to be sure that all the thicker paint in the bottom of the can is mixed well with the thinner paint in the upper levels. Repeat this mixing and stirring process every day when you start work and after any prolonged interruptions.

You will be able to use a roller for the wide expanses of the ceiling, but you will need a brush for the detailed work. The roller is much faster and easier than the brush, but be advised that rollers use considerably more paint than brushes, and no matter how careful you are, there will be a fine spray from the roller that will fly across the room and speckle everything in its path.

The brush is slower and often leaves brush marks that are evident long after the paint has dried. When you are working overhead, you cannot load your brush fully or the paint will run through the bristles and down the handle and into your hand.

When you decide to paint a ceiling, remember that you do not have to make it all the same color. The ceiling can be stenciled, designed geometrically, and have dozens of interesting facets to the color scheme. Many ceilings have as many as five or six colors in them. Many start with a bold band of color a foot wide, then blend with a muted shade or pastels, and conclude with a design in the center. Decorating a ceiling can be as adventurous as decorating a wall.

WALLPAPER AND CARPET AS CEILING COVERS

In recent years many homeowners are using wallpaper and carpet fabrics as ceiling covering. To apply wallpaper, use exactly the same techniques you use when applying the paper on a wall. You will need a ladder or two and a long roller, in addition to the other equipment normally required. The major difference between ceiling and wall is a significant one: gravity.

When you wet the paper and start to apply it, normally the weight of the paper pulls it down the wall. When applying it on the ceiling you will have to pull the paper tight, after the top part of the roll has been applied. As you move from one corner toward the other, roll the paper firmly and be alert for air bubbles or wrinkles.

When figuring out the amount of ceiling covering to buy, multiply the length by the width of the ceiling to get the square footage. On many paint containers there is a manufacturer's message about the number of square feet the contents of the can will cover. For example, if your ceiling is 15 by 15, the square footage is 225. If the contents of a small container of paint will cover 100 square feet, you will need to buy three cans. If a roll of wallpaper will cover 16 square feet, you will need to buy 14 rolls. If you are hanging ceiling tiles, figure that some tiles will cover 2 square feet and others will cover 4 square feet. If a box of tiles contains 100 tiles 2 square feet each, one box will not be enough; you will need 113 tiles.

Your problem here is one of waste. Unless you have a real need for the 87 tiles that will be left over from the second box of 100, you have paid a needlessly high price for the tiles. When you run into such problems, ask the dealer if he has a broken box of tiles so that you can buy the extra 13.

Many dealers sell the 4-square-foot tiles individually. For the ceiling described above you will need 57 tiles. Do not buy a box of 100 if you can help it. Buy tiles separately if possible, and if not try to buy a partial box. When you buy a great deal more than you need, all the money you saved by doing the work yourself is negated.

Chapter **6**

Simplified roofing processes

A house is no better than its roof. A good roof is a part of a house that is seldom really noticed. A bad roof is the center of attention every time the weather turns bad.

The function of a roof is not merely to protect the house from rain and other forms of precipitation; it is vital to the heating processes as well. A good roof that is well insulated can save you a great deal of money on heating costs over a period of years. Heat rises, and all the heat in your house will make an effort to escape through the roof.

Heat rises first to the ceiling, where it is trapped unless it can find a way to permeate the ceiling. It banks there like an invisible cloud and as more heat accumulates, the heat gradually works down to the floor level. If there are cracks or heat leaks in the ceiling tiles or in the spaces between walls and ceiling, the heat will escape either into the attic or into the upstairs rooms, where the same process occurs.

You have probably noticed how warm the attic is compared to the rest of the house, or how much easier it is to heat the upstairs rooms than the basement rooms. If your attic is not significantly warmer than the rest of the house on a typical day, either your roof is allowing too much heat to escape or your ceilings are extremely heat efficient.

Check how well your roof insulates by observing it on a day when snow has fallen, if you have snow in your part of the country. Notice how quickly the snow melts from your roof compared to other roofs in the neighborhood. Notice particularly how fast the snow melts on unheated buildings such as toolsheds, garages, or barns. If the snow melts on your house quickly, you probably need to do some insulating. If the snow melts on the unheated buildings first, you have an exceptional roof that is well insulated.

A roof that leaks is a problem that must be attended to. A tiny hole can cause leaks that can ruin furniture, books, and walls. What is worse is that the entering water can sometimes leak inside the walls, where it might come in contact with exposed electrical wires or where it can rot sole plates and subflooring and become an invitation to roaches and other household pests that thrive in damp areas. Termites are especially attracted to dampness.

A good roof provides considerable protection from excessive heat in summer as well as saving heat in winter. A poorly constructed roof can cause the temperature inside a house to rise by as much as 20 degrees on a hot day.

If your house is under construction and you are ready to roof it, you have a fairly wide selection of roofing materials. These include metal, wood shingles, and asphalt shingles.

METAL ROOFING

One of the most important considerations in selecting roofing material is fire protection. Some insurance companies allow decreased premium rates if the roof is metal. The sheet metal roof has the definite advantage of being fireproof, and many home owners select the metal roofing for that reason alone.

One advantage of metal roofing has nothing to do with economy or fire safety. Some people find the sound of rain on a tin roof to be one of the most comforting and relaxing sounds in our environment. At the same time sleet or hail on a metal roof can be distracting.

A metal roof is easy to install and will last for decades. The only real maintenance problems are keeping paint on the metal and occasionally making repairs when the wind damages the metal sheets.

Installation

Metal roofing usually comes in sheets that are 2 feet wide and 8, 10, 12, or even 16 feet long. The 8-foot sheet is perhaps the most common.

Before you begin work, check the roof sheathing to see that it is properly prepared for roofing. Sheathing of modern houses makes use of some form of plywood or occasionally sheathing boards, which are 5-inch boards nailed at right angles to the rafters. Sheathing should reach from the outermost edge of the rafters, including covering the eaves area, to the midpoint of the rafter 4 feet away. The edge of the plywood should be located so that the panel that abuts it has nailing room on the top of the same rafter.

Frequently roofing paper is nailed over the plywood. Some builders like to include this step, others find it lacking in economy and troublesome because the first high wind that comes along will tear the building paper off and the job has to be done again. However, the building paper helps with insulation and also keeps the sheathing dry while you are waiting to install the roofing.

Start at the bottom of the roof slope or gable when you are ready to apply the metal sheets. The end of the metal sheet should extend beyond the rafter ends by about 2 to 4 inches. There are special roofing nails you can buy that have very large heads. Some of these roofing nails also have tiny circles or washers made of rubber or similar materials. When the nail is driven, the sealer material is flattened so that it seals the hole made by the nail, thereby preventing leaks.

Each sheet of roofing will have ridges at each side. Many roofers like to drive the nails in at these ridges because of the natural watershed properties of the ridge. Rain water cannot stand on the ridges and therefore cannot seep into the attic. Drive a roofing nail every 2 feet along the ridges as you install the first sheet, which should be positioned so that the end hangs over the rafters slightly and the extreme edge hangs over the eaves rafters slightly.

When the first sheet is installed, nail in the sheet that fits beside it. You will notice that the ridges are manufactured so that they fit snugly over one another. This overlap prevents any water from seeping or blowing under the edges of the roofing sheets. Lap the sheets and drive roofing nails through the ridges and into the sheathing. Continue across the entire lower part of the roof.

When you start the second row, allow the sheet to lap over the first sheet installed by about 1 foot. Measure the distance to the peak of the roof, and if the distance is such that you can do so without wasting roofing, lap even more. If you are using 8-foot sheets and the distance from the top edge of the first row to the peak is 14 feet, consider that you have 16 feet of roofing length for that part of the roof, which means that you have 2 feet of surplus length. Use the surplus for lapping. If you have only 12 feet left, you have 4 feet of surplus, and you can lap each sheet by 2 feet. The more you lap the less chance there is that rain will blow up under the sheets and leak into the attic (FIG. 6-1).

In rainstorms with very high winds the rain can be blown at an angle and with such force that when it hits the roof, some of it will be forced under the lower edge of the roofing sheet. If the force of the wind is great enough, it can drive the rain several inches up under the edge. Lapping, along with plenty of nails, is your best protection.

The question is how many nails to use. The more you use, the stronger the roofing will be. At the same time each nail makes another hole in the roof. You have to decide with reference to the kinds of weather you commonly experience in your part of the country if an unusually large number of nails will be necessary.

When the second row is completed, install the third in exactly the same manner. When you reach the peak, the upper edge of the metal sheet should be even with the top edge of the roof sheathing.

Next, roof the other side of the slope. Work in the same pattern, and when you have brought the metal sheets to the peak of the roof, you are finished this part of the job, except for the roof cap.

Adding the roof cap

The *roof cap* is a length of metal that is shaped to fit over the peak and lap the edges of the sheets that end at the peak. This cap is nailed tightly over the peak so that rain can't be blown under the peak.

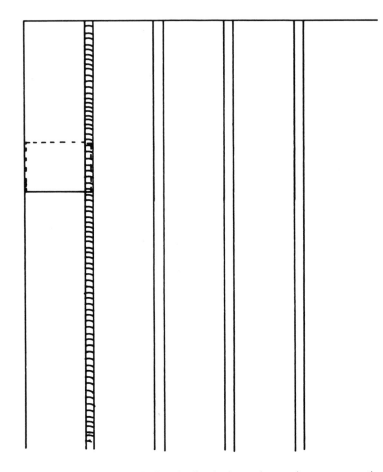

6-1   When installing a metal roof, allow the fitted edges to lap snugly over one another. Lap the bottom and top ends of the panels to prevent water from blowing in.

If a cap cannot be located, there is a substitute procedure: Use tin snips or shears to cut a section of flat metal from between the ridges. Then lay this strip lengthwise across the peak and bend it over the peak so that it can be nailed in place. Unless you keep the strip perfectly in place it can be less than attractive. The manufactured caps look better and work equally well, so it is best to use them (FIG. 6-2).

NOTE: There is a ridge in the center of the sheet of roofing. You can nail through this ridge as well as those on the edges.

Some roofers like to finish their work by applying tar or other roofing sealer over each nail. This final step is one major way to prevent leaking.

Do not try to install metal roofing on an extremely windy day. The sheets of metal offer a great wind resistance and you will be exposed to injury. The edges of the metal sheets can cut you if you are not careful, so use gloves whenever you are handling the material.

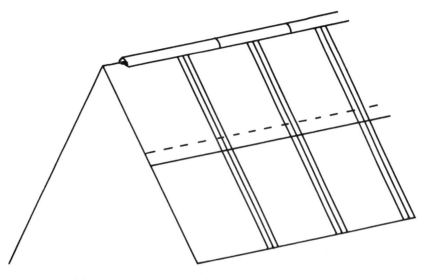

6-2 Metal capping sections fit over the peak of a metal roof.

SHINGLES

Many years ago most building codes rendered wood shingles unsuitable for home construction. The argument was that the dried wood was a perfect place for a fire to start if a spark from the chimney landed on the tinder-dry material. A collapsing roof could greatly endanger the lives of all those inside the house.

Modern technology has produced a fire-retardant wood shingle that can pass many if not all building codes. These shingles have all the charm and attractiveness of the old wood shingles but they have the built-in safety factor that is designed to prevent roof fires.

Installation

As with any roofing, the first step is to determine that the roof sheathing has been properly installed. When you are ready to begin the actual work, install the first shingle in the corner of the roof slope so that part of it overhangs the eaves on both roof edges at that corner. Nail the first shingle in all four corners but stay at least an inch or more from the actual corner to avoid the danger of splitting.

Install the second shingle so that it abuts the first. Some roofers point out that you need to leave space between the shingles to allow for expansion and contraction. If you choose to leave this space, there is no need to make it any larger than the thickness of a quarter. Complete the first row of shingles in this manner.

After the first row is completed, return to the same row and install shingles so that they lap the cracks between the shingles in the first row. By doing this you close any leaks that would be present at the start of your work (FIG. 6-3).

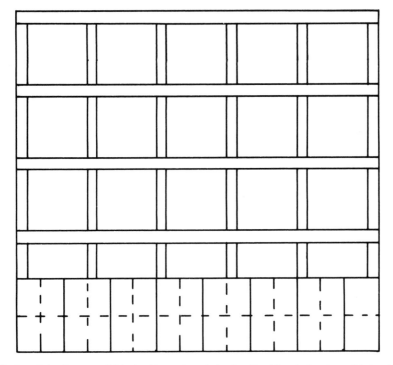

6-3 Install the first row of shingles (shown here in broken lines) by placing them side by side. When the first row is installed, return and nail in a second row so that all spaces are covered. In succeeding rows always lap the cracks between shingles.

The second row should lap one-half of the first row. This might seem like a waste of shingles, but it is necessary. There will be a potential leak in the crack between any two shingles unless you overlap the shingles to form a perfect watershed. You will need to lap all shingles halfway throughout the remainder of the roofing job.

Proceed along the second row and nail in all the shingles. Then move to the third row, remember to lap halfway, and cover that row as you did the first two. Remember to keep nails at least an inch from the edges of shingles (FIG. 6-4).

Planning the peak

At the peak will come the only major difficulty of the operation. The time-honored method of installing wood shingles is to determine first where the predominant winds originate in your area and to design the roof accordingly. If most rains and other storms blow in from the north, then the higher shingle course should be on the north side of the peak, if your house runs in a north-south direction.

If you are located in an east-west direction, determine whether you are more likely to receive the majority of storms from the east or west. If you decide that west is the most likely direction, then on the east side of

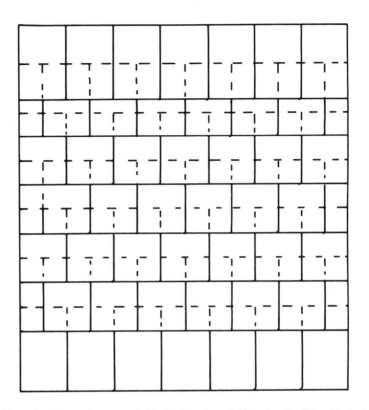

6-4 Always lap the previous row of shingles by at least half the length of the shingle. In this way every part of the roof that could admit water is covered.

the house install the shingles until they reach the exact peak. Nail the last row at the top and at the bottom.

When you reach the peak on the west side of the house, install the next-to-last row of shingles so that they reach the peak level. Then install another row of shingles and allow these to stick up above the peak by at least the length of half a shingle or a bare minimum of 6 inches (FIG. 6-5).

The purpose of this step is so that the shingles will slant over the peak and any rains that fall straight down will hit the shingles before the raindrops have a chance to seep in at the peak. When the moisture hits the shingles it will be deflected down the west side. The shingles on the east side abut the ones on the west side, and even rains from the east will have a difficult time penetrating the water barrier created by the shingle arrangement.

If you installed the building paper under the shingles and capped the peak with the same paper, any rain or other moisture that seeps past the shingles will run down the surface of the building paper and off the eaves.

Wood shingles form a remarkably leakproof roof. Even if there is no building paper and you omitted the sheathing, and even if you can see daylight through the roof, the shingle laps will prevent nearly all types of

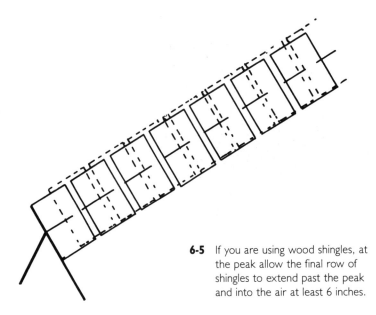

6-5 If you are using wood shingles, at the peak allow the final row of shingles to extend past the peak and into the air at least 6 inches.

leaks. The watershed installation deflects all water except for an occasional drip that finds the precise opening to penetrate the roof.

When you use the sheathing as well as the building paper, you are not likely to have one drop of water penetrate the roof. When or if shingles are broken or blown off, you can slide a replacement shingle under the lap and nail the new one in place without disturbing the remainder of the roof arrangement.

ASPHALT SHINGLES

Asphalt shingles are not similar to wood shingles except that they are installed with basically the same types of overlapped layers of water-shedding materials. They are easy to install and fairly economical.

Calculating need

You buy asphalt shingles in a unit called a square. A square will cover 100 square feet of roof space. To determine how many squares you need to buy, multiply the length by the width of the roof and divide by 100. For example, if the house is 52 feet long and 32 feet wide, your roof square footage will be 1664.

Note that the addition of about 10 percent for waste means another two squares of roofing.

Note that a 1-inch overhang keeps water that drips from your roof from running down the sides of your house. Otherwise, runoff will wash down the siding and will discolor it. Water will also drip over windows and will splash on the storm windows and soil them with the dirt that is washed off the roof. Occasionally the water will even drip past the window framing and into the window itself, where it can find its way into the interior of the walls and cause decay and encourage insect infestation.

You also will need extra shingles with which to make a cap for the peak (FIG. 6-6). Therefore, you will be economically justified in buying an extra square.

6-6 The more roof angles there are, the greater the importance of eaves or overhangs. A roof like this one can easily be made waterproof by fitting shingles carefully, using overhangs, and sealing the edges where shingles fit against the siding.

Installation

Start installing shingles along the edge of the roof line at the overhang. Let the first shingle lap over the eaves by 1 inch. This lap will protect the eaves from water decay. Also, let the shingle lap over the edge of the roof slope side by 1 inch. Nail in the shingle by driving a roofing nail near the edges of the midpoint of the shingle, just above the split. Drive two more nails at the upper corners.

One important point: turn the first row of shingles upside down so that the split, which is normally at the bottom, is now at the top. Install the entire first row in this fashion. When you have finished, install another row of shingles over the first one, but turn these the correct way. The purpose of the upside-down installation is to prevent the splits from aligning, and so that the adhesive on the back of the first row will be in contact with the roof.

Abut the second shingle to the side of the first, but let the end lap over the eaves as you did with the first one. Continue in this manner across the roof line.

Lap the second row of shingles halfway, so that the split is almost aligned with the shingle below it. Nail each shingle in place as you go. As you start a row, you might need to cut a shingle so that half of it is installed. Cut from the split to the top of the shingle. The purpose of this is so that the split lines will not ever be aligned anywhere in the roof— that is, no two successive shingles will be spaced exactly alike (FIG. 6-7).

Continue with the installation until you reach the peak. Then go to the bottom of the opposite slope and start the process there. Work until your two roofed slopes meet at the peak. Here is where the cap will be installed.

Turn the shingles sideways and fit them over the peak. Some roofers prefer to cut the split portion of the shingle off and discard it, using only the solid part. Others use the entire shingle but start by doubling the first one in the cap row, just as described above for the first row of regular shingles.

Proceed to cap the entire peak. The solid part of the shingle covers the split in the one below it, so there is never a point where water can find a direct entrance.

SPECIAL PROBLEMS

In any roof there is likely to be some small difficulty—unless it is a shed roof, which is by far the simplest and most economical roof you can install. The shed roof simply starts high at the front and concludes low at the back, with a drop of at least 1 inch every 5 feet, and often with a much greater drop. In a shed roof that is 20 feet long there should be a drop of at least 1 foot to provide the needed watershed.

One point to remember is that the higher the roof—the greater the slope—the better the roof will shed water and snow. This can be a serious consideration if you live in a cold climate. An accumulation of several feet of snow can put severe strain on a roof, and unless you have very strong

6-7 Even a roof with a gentle slope like this can be waterproof if shingles are spaced properly. On gentle slopes there is a greater need for sealing the roof and siding edges.

rafters and thick sheathing, you might experience a collapsed roof or one that begins to sag. This is particularly important if the roofing timbers were not properly dried before installation (FIG. 6-8).

Many building codes require that the wood contain no more than 19 percent moisture content. Most people have no way of measuring moisture content and do not know if the wood was properly cured. If the rafters are installed green, there is a good chance that they will sag and bend under the weight of snow. If the roof is composed of superior materials, the chances of this happening are greatly reduced.

6-8 In more complex roof systems, the roof must be well supported and watertight.

Although the slope of the roof is helpful in shedding water and snow, higher slopes provide greater wind resistance and so heighten the possibility of damage during extremely high winds. Inland winds do not reach great speeds in most parts of the country, but in mountainous regions or on plains there is a chance of gale force winds or higher. Under normal weather conditions here is the general rule of thumb: In areas with heavy snowfall roofs should be steeper; in areas with high winds roofs should have less slope.

Although a shed roof is the most economical type to construct, it will not be suitable for most houses. The shed roof is best suited for buildings with four rooms or footage of 600 to 1000 square feet. For houses with square footage of 1500 square feet or so, the typical roof used is the gable roof, one that has two slopes and a peak in the middle. The gable roof is considered the second most economical and efficient type of roof.

When dormers or additional gables are constructed, the roofing problem becomes slightly more complicated. In the valleys between the normal roof slope and the gable, there will be a rush of water as it is shed by the two slopes and diverted into the valley. Because of this excessive water, there is a greater tendency for the roof to leak or experience damage. In these valleys you will need to use metal flashing and/or caulk to seal the shingle juncture well. Install the shingles on the slopes at dormers or gables as though the slope were the full roof. That is, start at the bottom of the slope as you did for the normal roof slope. You will need to trim shingle edges as you cover the angled slope. Caulk generously in these valleys and make certain that all leak possibilities are carefully sealed (FIG. 6-9).

At dormers you will need to shingle the roof just as you would if the dormer were the entire building. The dormer will be an offset in the roof line, and while the regular roof shingles run in an east-west direction, dormer shingles will run in a north-south direction. Start dormer shingles at the lowest edge of one slope and install the first shingles upside down as you did before. Finish roofing both slopes and then add the peak cap for final weatherproofing.

GUTTERING

The major purpose of guttering is to divert all of the water that drips off a house during a storm to one or two main areas. You have noted before that water that constantly drips down the side of a house can cause unsightly appearances and damage, and guttering is the best way to control the amount of water that affects the sides of the house.

The guttering materials available today are much lighter, more workable, and in general more satisfactory than the heavy guttering of years past. Much of the guttering today is aluminum, a good metal with which to work.

Because of its lightness and thinness, aluminum is easy to handle. Most aluminum is hung very simply. The top of the gutter trough has a turned-back ridge, and just under this ridge is the best place to install the long metal spikes known as ferrells, if you must insert the spike through the metal. This can be done with ease because of the lightness of aluminum. Some guttering is equipped with bands or supports that fit under the gutter trough and extend high enough that the spikes can be run through the prepared holes and into the eaves framing of the house.

When you install your own guttering, you will need to measure the length and width of the house and buy sufficient units to cover all of the roof area. In addition you will need to buy the couplings, ells, and other

6-9 A roof with many angles and slants can be more complicated, but also among the most attractive possible.

necessaries such as downspouts and clamps. Usually you can ask a salesman to help you with what you need. Nearly every new house that is built will have guttering, and while much of it is put up by special crews, a great deal of the material is still sold over the counter.

The easiest alternative is to have guttering installed by a firm that specializes in such work. They will have equipment that molds the guttering on the site. The fit will be neat and the workmanship, in most cases, flawless. Saving the installation cost is the only argument in favor of doing this job yourself.

Calculating need

Your first step is to get an accurate measurement of the roof line at the edge of the slope and along the eaves. If the roof is a standard gabled house with one peak and two slopes and no offsets, you will need to have two basic dimensions. If the roof line is interrupted at one or more places you will need to measure all of the roof drain distance around the offsets.

If the house roof line is 56 feet on the long side this will be your drain side. You will need 112 feet of guttering. The new aluminum guttering and vinyl guttering will usually come in lengths of 10 feet. You will need to buy 12 units of guttering (the basic trough) plus one extra length for the 2 extra feet. The typical cost of one unit of guttering trough is about $7, so your investment at this point will be around $91.

You will also need quite a lot of additional materials. The end caps are units with a closed end that are installed at the end of each trough. You will probably need four of these at $1 each. Each of the 10-foot sections of guttering trough will have to be connected. You will need 13 of these, one for each unit of trough you bought. They each cost about $1.30. At this point your cost is slightly more than $111.

Your guttering trough length is 112 feet, and you will need a support bracket every 2 feet. You will then need 56 of these at about $1.25 each, for a total of $75. Corner joints sell for approximately $4, and you will need four of these.

You must decide where to locate downspouts. Many people like to have one on each corner. Others who wish to divert the flow of water away from certain areas of the house and yard may install them on one side of the house only. If you decide to have four downspouts you will need four connectors. These sell for about $5 each. You will also need four elbows if you choose to install four downspouts. These elbows act similar to universal joints in that they can be turned to provide the exact angle needed to connect the guttering trough to the downspouts. Two elbows per downspout are needed, and these cost roughly $1.75 each.

You will need next the downspouts. These come in 10-foot sections and you will need at least four of them if you have four drains. If your eaves height is more than 10 feet, you will need two per drain at a total cost of $80 for all four drains. Downspouts must be joined, and you will need one joiner for each juncture of two downspouts, or four for the entire house, unless your house is a two-story structure, which requires more downspouts. The cost here is $6.40. You will also need a down-spout holder or bracket, cement, and screws or spikes for mounting.

The remainder of your cost will bring the total expense of guttering materials to about $330. Prices vary from one part of the country to another; these prices represent a nationwide average.

Before you make purchases and begin to work, call a guttering firm to have them provide you with a free estimate. If their price is close to the one provided here, hire them to do the job. If you can save considerable money by doing the work yourself and want to do so, buy the materials and start installing.

Installation

Guttering used to be installed with spikes or ferrells, which could be very difficult to work with. The job required you to drive the ferrell through the outside guttering trough wall, extend it to the inside wall, drive the point through the inside wall, and then drive the spike into the eaves facing. The problem was that damage often occurred to the guttering trough or spike, or the facing of the eaves. This type of guttering material is still sold. However, the newer type that uses screws and brackets is much easier to install and is a much better choice. There is a good chance that a dealer who sells the spike-installed material is still holding outdated material, although this is not always the case.

The steps for installing screw-and-bracket guttering, are simple. First, use the screws provided and install the brackets into the eaves facing. Space them every 2 feet. Tighten screws firmly, being careful not to over-tighten them. This could cause stripped wood fibers, which would prevent the screws from holding.

When all brackets or mounts are installed, you can insert the trough sections. These snap into the brackets quite easily. Do not force them.

Everywhere you join two units, use a cement or glue. The cement is not needed as much to hold the units together as it is to seal the joints well. If joints are not sealed, they will leak badly and you will still have water dripping down the side of your house.

Proceed to the end of the roof line. At this point you can install the center drop into which your downspout will fit. At the end of the trough line install an end cap to keep water from flowing past the downspout opening and out the end of the trough and down the corner of the house. Fit the downspout into the center drop and install holders or brackets along the corner of the house and attach the downspout to these. Cement all joints where there is a need for a water seal.

When you have completed the work, choose whether to paint the guttering or leave it as it came from the manufacturer. Most new vinyl guttering never needs to be painted unless you want to try to match the colors of the house. Check with your dealer or paint store for the best paints to use on the particular guttering you have installed.

If you have large trees growing near the house, you will need to clean out the gutters regularly after each leaf fall. Twigs and dead leaves can clog the guttering and cause an overflow. One of the worst aspects of an overflow is that the decaying leaves and twigs will disintegrate and the tiny specks of dirt will be deposited along the side of the house as the water carries the debris downward. As more and more leaves accumulate, the gutters will become heavier and heavier. Grit from the roof shingles will also wash down and these will be added to the overall weight. When excessive weight accumulates, the brackets holding the guttering can bend or the troughs can warp.

The next question is what to do with the water that is diverted into the downspouts as it emerges onto the ground below. Many people add an extension to the downspout and channel the water away from the house, and onto the grass. This is acceptable if the soil around the house has been graded so that it carries surface water away from the house, although for several hours after a heavy rain the area will be soggy and damp.

Diverting water from guttering

One of the best methods of diverting water is to install underground flow pipes. You can bury a pipe or drain tile then connect the end of the pipe to the end of the downspout. In this manner water can be directed to a nearby garden or other acceptable part of the yard.

Drain tile is a length of very flexible tile with holes spaced throughout the top side. This is the same material that is used around the outside of basement walls to carry seeping water away from the basement. The holes in the top allow the water to enter and the bottom of the tile is solid so the water will flow in the desired direction. Bury the drain tile so that when the water level reaches the desired height it will overflow through the holes and water the lawn from under ground. You can also create much the same effect by using regular plastic pipe with nail holes punched in it so that the pipe will leak just enough to keep the roots of grass or other desired plants moist but not saturated.

ROOFING MAINTENANCE

When all of your roofing work is done, you will still have ordinary maintenance duties to perform. While these are simple and usually effortless, they are necessary. Inspect the roof after severe wind storms to see if shingles have been loosened or blown away. If you have a metal roof, look for corners that have been loosened and bent upward. When you see a problem, repair it immediately. Renail the corners of the metal sheets or replace the shingles that have been blown off. When guttering downspouts are not carrying the expected load of water, check to see if there is an overflow somewhere along the troughs. Make periodic visual examinations of the roof throughout the year.

When you discover a leak in the roof, set a wide container under the suspected leak. During the next rain check to see if water is dripping into the pan. If so, examine the roof to see if you can spot the point of entry. If not, check rafters or braces to see if water is entering at one point, running along the wood, and dripping into the container. Often a leak is 8 to 10 feet from the point where the drip is found. When you find the leak, push a nail point upward through the spot to mark it. Then when the weather is dry, locate the nail point and make the repairs by putting a spot of tar on the leaky spot.

One final suggestion for roofing concerns installing insulation on top of the sheathing. Normally you nail the sheathing over the rafters and then lay insulation between the ceiling joists, or you staple it between the

6-10 Lay insulation between the 2 × 4s, then cover with more sheathing and shingles. Such a roof is more expensive than the traditional type, but the strength, waterproofing, and insulation value are so great that the roof is actually more economical in the long run.

rafters but under the sheathing. You can also install insulation on top of the sheathing. Make rectangles that are 4 × 8 feet by nailing together 2-×-4 lumber. Then lay the batts of insulation inside the rectangles. For best results, stand the 2-×-4s on edge as you nail them in place. When the 2 × 4s are ready, lay the batts inside and then cover the rectangles with other sheathing before adding your roof. This process takes time and it is relatively costly, but you will have a roof that has great insulation capacity and the investment will be worthwhile (FIG. 6-10).

Chapter **7**

Finishing home additions

*F*inish carpentry involves the final touches you put on a house. The term applies to the building of a new house as well as remodeling or adding on to an existing house.

When you add a room to your house, you want the new room to have an individualistic charm. Do not worry if the new room creates a radically different appearance from that created by the older part of the house. You can help soften the impact of the new room by making a few subtle changes in other parts of the house that will make the new room seem like a proper extension of the old.

One of the most noticeable things in a room is the facing or framing of the door, in addition to the door itself. If the older part of the house is Contemporary American and the new room is Early American, start to blend the two rooms at the doorway. For example, install a sturdy door that is painted a solid and bold color on one side and a deep brown on the opposite side. The door framing on one side can follow the contemporary style and on the back side it can be made of thicker and heavier wood with a stain rather than paint. These contrasts tone down the changes from one room into the other.

The walls of the new room can also be a statement of the time and mood represented. Wainscoting and period wallpaper can be used. Darker walls, heavier drapes, and ponderous furniture all suggest antiquity and set the mood of the room.

While molding around the ceiling in one room may be thin or even nonexistent, in the Early American room the molding can be very thick and heavy and stained a dark brown—nearly black—to blend with the darker hues of the walls.

Floors can be made of wide boards and covered partially by braided rugs rather than wall-to-wall carpet. Light fixtures can be muted and heavier, with the patina bronze effect. Lantern or candle simulations also suggest an earlier age.

As you see, much of the room's statement is accomplished by furnishings, but various aspects of finish carpentry can go a long way toward achieving the desired effect. Choice of materials is of great importance, and you will find that many heavier materials cost considerably more than the lighter, brighter contemporary counterparts—although this is not always the case.

PINE FLOORING

When you start to floor a room, your first thoughts often turn to carpet or tile. Also consider wood flooring.

Traditionally oak, maple, or similar hardwoods were the wood of choice for flooring. Pine, however, is a superior wood to work with and has been the victim of unfair criticism during the past decades. The general impression of pine is that it is a soft wood that splinters easily and will not endure. Cedar has the opposite reputation; it is considered one of the hardest and longest-lasting woods available.

There is much confusion concerning hard and soft woods. To the lumber man a hardwood tree is one that is deciduous—that is, it loses its leaves in the winter—while a softwood tree is one that keeps its foliage and has needles rather than wide leaves. Such generalities, however, can be misleading.

For example, poplar is considered a hardwood, yet it is one of the softest woods when it is green. When properly seasoned it becomes one of the hardest. The same is true with pines. Heart pine is much harder than the heart of a poplar tree; the heart of the poplar is in reality the softest part of the tree and the part that decays first, while the heart of the pine can be incredibly hard even after 100 years of use in a house.

For a floor that is beautiful and highly serviceable, thick pine boards are a good choice. You will also be surprised at the relatively low cost of such a floor (FIG. 7-1).

Calculating need

If you have a room 16 feet long and 14 feet wide, you can buy pine timbers that are 2 inches thick, 10 inches wide, and 16 feet long for about $10 each. For about $150 you can have a floor that is a full 2 inches thick and so sturdy that it will never suffer from wear. Such a floor, if properly treated and cared for, can last for over a 100 years. If you want to settle for boards that are 1 inch thick, you can save considerable money.

An oak floor of the same square footage will cost you about $360 or slightly more, if the flooring is $2^{1}/4$ inches wide and $3/4$ inch thick. Lengths vary from 2 or 3 feet up to 5 or 6 feet.

With these wood floors subflooring is not necessary. If you choose to install subflooring in addition, your floors will be so thick that they will not creak or give no matter what kinds of household weight is put upon them. You also won't need to spend a lot of money on tools for installation: All that is required is a saw, hammer, nails, and a few other basics.

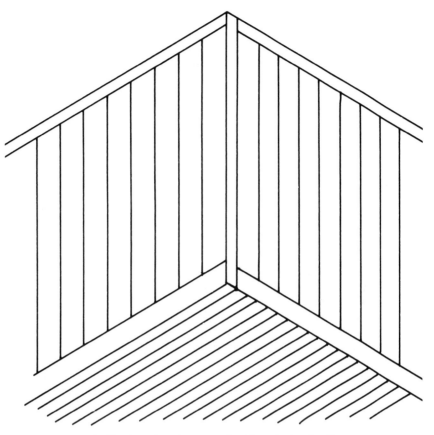

7-1   Pine boards make a very durable and attractive floor.

Installation

To begin, push the first board firmly against the far wall so that the edge of the board fits flush against the wall. Remove molding before you start to install the flooring boards. With the board in place, drive 16-penny nails along the edge that is near the wall. Drive the first nail 1 foot from the end of the timber and space the others 2 feet apart. Keep all nails close to the wall but no closer than 1/2 inch to the edge of the timber. If you drive too close to the edge of virtually any timber you run the risk of splitting it slightly.

When the inside edge is completed, you are ready to secure the outside edge. You need to decide whether you want nail heads to show. If you do, use larger nails with heads that are particularly impressive when painted black against a background of lighter wood. If you don't want them to show you have two basic options: The first is to toenail or angle-nail into the side of the beams. The second is to use a punch to drive the heads of the finish nails slightly below the surface of the wood. Then you can fill in with a plastic wood that will conceal the heads.

If you toenail the outside edges, you will still have to decide how to handle the inside edges of subsequent boards. The first board you installed could be nailed with the heads so close to the wall that the floor molding will cover it. You cannot do that with later boards, so you may decide to use the punch and hide these nail heads.

As you install each board, push it firmly against the previous one. Do not allow any cracks to show unless it is impossible to close them. To achieve this, start the nail and drive it nearly through the board. Then place a length of 2 × 4 against the side of the board to be installed and strike it sharply with the hammer, forcing it over as far as you can. Place your foot firmly against the board and hold it firmly while you complete the nailing.

Fitting the last board

When you reach the final board in the room, you will need to measure the existing space between the next-to-last board and the wall. If the space is not enough to allow the board to fit, you will have to use a circular saw and rip the board to the appropriate width. You can mark the cut line with a chalk line.

The room might not be perfectly square. Measure at one end of the room and write down the distance. Then measure at the other end and set down that mark. Compare the two. If they are not the same, you will need to chalk so that the differential is recognized and the proper cut can be made.

If one end of the room has a gap of 5 inches and the other end a gap of 7 inches, you will need to mark the board accordingly, remembering to angle the cut so that the angle of the room accepts the fit well. If the north end of the room is narrower than the south end, be sure that you make the cut accordingly.

When making the cut remember that floor molding will cover the last inch or more of the board, so it does not matter if there is a slightly bad fit. A loose fit is better than cutting the 16-foot board several times and taking the chance of ruining the board by trying to make the best possible cut.

When you are cutting a 2-inch board with a circular saw, the saw blade barely reaches through the board. You do not need to elevate the board greatly in order to make the cut. If you want to saw the board on bucks or sawhorses, put a 2 × 4 under the board, positioned so that the 2 × 4 will not be in the cut line. The height of the 2 × 4 will prevent your cutting the material under it.

Painting or staining

When the final board is in place, you might wish to cover the floor with some type of paint or stain. Stain works beautifully on wide pine boards. You can use stain that is an authentic pine color or opt for a lighter or redder color. You can use a darker stain to give the floor an oaken look.

Before making a decision consider how the stain will look with the furniture in the room.

An advantage of stain is that if you try a color and find that it is too light, you can apply a darker stain over the lighter one and gradually attain the shade your desire. If you get it too dark, a little sandpaper or stripper will remove the stain with relatively little effort.

You might find that the flooring boards are not perfectly uniform in thickness. If you have problems with uneven board edges when the boards are installed, put a sanding disk on your drill and run the sander over the edges for a short time and they will be so nearly even that you cannot tell simply by looking that there is a problem.

HOMEMADE PARQUET FLOORING

An interesting variation on the type of floor described above is the parquet floor. You can make your own parquet floor pieces in only a few minutes and with a minimum of effort.

Preparing the parquet

Begin by measuring the exact width of a 10-inch board, which will be about 9¹/₂ inches. Then measure down that same distance on the board and mark the location. Use a square to mark a cut line across the board. Take care to keep the cut line perfectly square. When you cut along the line you will have a square section of flooring 2 inches thick and nearly 100 square inches in size. Continue cutting until you have a good stack of the sections.

Installation

Use your carpenter's square and search for the corner that is closest to square. Start installing your squares in that corner (FIG. 7-2).

Fit the first square snugly into the corner and nail it in place using finish nails. You can again use the punch to drive the nail heads below the surface and then apply plastic wood over the heads.

If plastic wood is too tedious for you in this work, scoop up a handful of sawdust and squeeze a small amount of glue into a small saucer or otherwise useless container. Mix the sawdust in with the glue. When it is thoroughly mixed, you can pack the sawdust into the holes above the nail heads and the mixture will hold and look like wood.

You can also toenail the squares in place (FIG. 7-3).

Continue to install squares down the wall, and when you are finished start on the next row.

Run the grain in opposite directions as you work. If one square has a north-south grain, turn the next one east-west so that the two grains offer contrast. Alternate lighter and darker blocks for contrast, or choose squares that are slightly different in color or wood grain.

For a good look, run a bead of glue along the edges of each board to

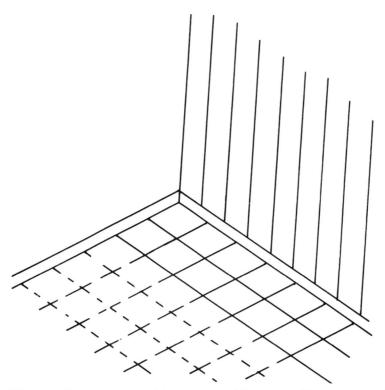

7-2   This type of floor can be created from the scraps left over from joist timbers or other work. Be sure to start with a square corner.

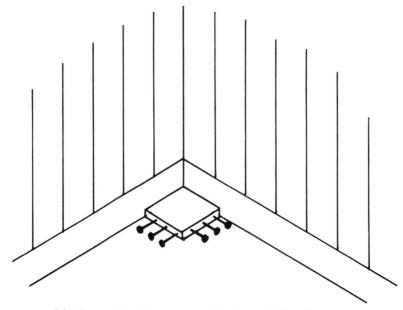

7-3   By toenailing the squares, nail heads are eliminated from sight.

be installed. The glue will keep the edges tightly pressed against each other and prevent dirt from falling into cracks between squares.

You can use any size squares for this type of floor, although wider boards are easier to work with and produce the most pleasing results. Achieve interesting results by mixing sizes of the wood. For example, try installing the 10-inch squares diagonally across the room in the center. Then in the open triangular spaces on each side of the row, install squares from an 8-inch board or 6-inch board. The squares will not fill the space or, if they do, deliberately leave enough space that you can install short sections of the ripped 2 × 4 to add a note of contrast to the design (FIG. 7-4).

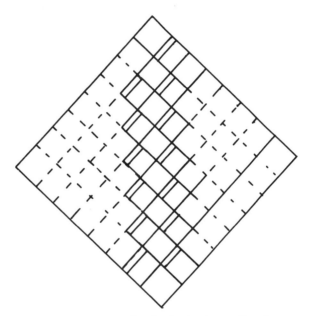

7-4 Create interesting variations on flooring by leaving small sections open and fitting 2- × -4 sections on edge into the spaces.

BRICK FLOORING

Lay a brick floor just as you would lay a patio. You need sturdy subflooring or concrete under the bricks so that there will not be any "give" when people walk over the floor. If there is a slight give, the mortar between the bricks might crack slightly.

Bricks, because of their slender size, offer dozens of very attractive possibilities in patterns. You can lay them in a very uniform manner, or you can mix horizontal rows with diagonal rows and you can have squares and diamonds spaced around the room (FIG. 7-5).

To install a brick floor, you will need several hundred bricks, preferably old, handmade ones if they can be found. Although these are irregular and difficult to keep even, they produce the most interesting designs in floor patterns.

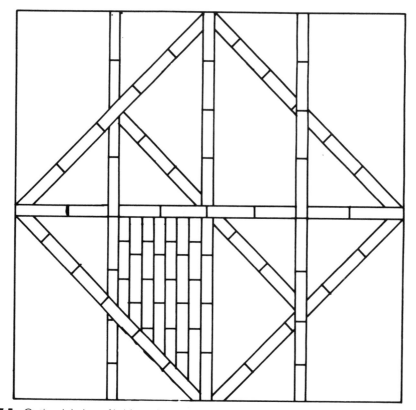

7-5 Optional designs of bricks make a more personalized floor. Use a variety of geometric designs and use bricks of different colors for more variety.

Installation

Start by mixing a boxful of very plastic mortar and spreading the mortar over the subflooring surface. You can use a sheet of plastic over the sub-flooring if you wish to protect it. The mortar should be 1/2 inch thick. Remember that you are, in effect, lowering your ceiling height the equivalent of the thickness of one brick stood on edge, so you will have to make adjustments for door frames and other aspects of the room.

Push the first brick into the mortar, then butter the next brick by applying mortar to the face of it. Do not put mortar on the edge. Push the new brick against the first one and if any mortar squeezes out, use the trowel to scrape it off and dump it back into the mortar box.

Keep the mortar joints even with the surface of the bricks. Try to position each brick so that it stands straight and level. A brick that is tilted will have one high and one low edge, either from side to side or from front to back.

If the height of the bricks standing on edge creates problems concerning door openings, you can use solid bricks and lay them flat. You will need fewer bricks this way but the pattern arrangement is somewhat limited.

An ideal place for a brick floor is in a basement, particularly in a room where there is a huge fireplace with a wide profile. The basement floor of concrete makes a perfect support for the weight of the bricks.

Even if you do not have a concrete floor in a basement, you can create a brick floor over dirt quite easily. Spread a base of sand, and over the sand spread 2 inches of gravel. Then spread a thick mortar base over the gravel. You will need to pay constant attention to the level of the bricks. You can use a board standing on edge as your level marker, or you can use a carpenter's level. The thicker the mortar base, the easier it is to keep the bricks level, because you can push them deeper, if necessary, into the mortar. Remember to use a sheet of plastic over the gravel.

TILE FLOORING

Ceramic tiles are a beautiful flooring choice. There are dozens of types and patterns of ceramic tiles from which to choose. Ceramic tiles are kiln fired so that they will retain their durability, hardness, and resistance to fading. Another advantage is that you can start the tile on the floor and continue up the walls. Be prepared, however, to pay a fairly high price for tiles.

The cost per square foot for ceramic tiles ranges from $2 to $3.50. If you have a floor that is 16 by 14 feet, you can expect to pay $450 to $800 to tile the floor. This price includes the tiles only, without any installation costs.

Special adhesive is used to lay ceramic tiles. This is available from the dealer who sold you the floor covering. Spread about 20 square inches of adhesive on the floor surface, after first thoroughly cleaning the floor. The surface must be free from grease and similar substances that would prevent the adhesive from bonding to the floor surface. When the adhesive is spread, lay the tiles in the same manner as a brick floor. Rake out the mortar joints as you desire.

PANELING BOARDS

You can buy sheets of paneling, which is one of the most popular wall coverings used today. You can also buy ordinary boards, fit them yourself, and save a considerable amount of money.

Thanks to modern milling techniques, you can buy boards in a variety of thicknesses and widths. Some of the advantages of these boards: They are highly attractive, they aid soundproofing and insulation, they are reasonably inexpensive to install, and they can actually strengthen the room framing considerably.

For best results, buy a mixture of boards that are 5, 8, and 10 inches wide and all 1 inch thick. Lengths should be the height of the room you plan to panel.

Using nailing strips

When you start with a fully framed wall, nearly all of the stud work runs vertically. You have very little space where you can nail a vertical board, so you might need to install nailing strips. These strips can be 1-×-4 boards, four of them per wall, that are nailed to the existing framing work. The first board should be nailed against the ceiling, the second one-fourth of the way down, the third one-half way down, and the fourth against the floor.

These boards will provide great support against any pressure put on the wall. When you push against a wall and feel a certain amount of "give," the impression is that the wall is weak and poorly constructed. When you push against a wall that has firm support, the impression is that of a solid, strong wall. These four boards can make a world of difference not only in how strong the wall appears to be but how strong the wall actually is.

Nail the cross boards to the corner framing. Be sure to use nails that are large enough to hold the boards securely and that the boards are firmly nailed in place.

Installation

When you are ready to install the first paneling board, square it to fit the corner, if necessary. This means getting a square start. If the corner is not perfectly square, the board will tilt when it is installed, and every successive board will tilt similarly.

To prevent this problem, set the board in position after cutting it to length. Each board should be carefully measured and cut only slightly shorter than the distance from floor to ceiling. If the ceiling is exactly 8 feet high, cut the board 7 feet, 11 3/4 inches long. This slight shortage will allow the board to fit easily into place. If you find that the thickness of the board causes the end to scrape the ceiling, take it down and cut off a little more.

Set the board into place and note the way the board fits against the corner framing. If there is a discernible gap between the board and the framing, measure the gap carefully at the widest point. Assume that the board fits well at the top, but the bottom is 1 inch from the framing. Remove the board and lay it on a flat work surface. Mark a point 1 inch from the inside corner of the board and then draw a line diagonally from the point to the bottom corner of the board. This is your cut line. You are, in essence, fitting the board exactly by shaping the edge to conform to the corner.

When you make the cut, start at the top end and keep the guide on the saw exactly on the line at all times. Saw at a steady speed and follow the cut line until it meets the corner. Now the board should fit well into the wall corner.

Nail the board in place by driving one nail partially at the inside top of the board. Then swing the bottom of the board firmly into place. Start another nail at the outside bottom. When this is done, put the level against the outside edge of the board to check for vertical trueness.

If the position is correct, finish driving in the nails and add the nails that are needed to secure the board. You should install two for each of the support boards for a total of eight nails. The board should now be very firm and strong.

Set the next board in place and fit it snugly against the first board. After starting two nails to hold the board in place, check it for vertical trueness. If the reading is good, install the board as you did the first one.

Vary the boards according to widths. You might want to try a narrow board in the corner, a 10-inch board next, then an 8-inch board, then a 5-inch, followed by a 10-inch board. Do not try to follow a rigid pattern of arrangement unless you specifically want a symmetrical wall.

When you are three-fourths of the way across the wall, measure the remaining space. See how many boards will fit into it without having to saw one. If the space left is 48 inches, consider that you can use four 10-inch boards—which are actually nine inches wide—for a total of 36 inches. You have 12 inches of space left. You can use two 6-inch boards and have a small amount of space left. This space will be covered by the corner molding. Or you can use three 10-inch boards (for a total of 27 inches of wall space covered) and three 8-inch boards (for a total of 21 inches covered). You can alternate 10-inch and 8-inch boards and the space will be filled neatly.

Cover the other walls in the same fashion. On the second wall you can abut the edge of the first board to the edge of the first board installed. You can cover all the remaining walls quickly and easily.

When you come to windows and doors, either fit the board panels against the framing or take down the framing and install the boards underneath. One word of caution: When you install board paneling under existing window frames, the door frame and the window frames extend into the room, which can create problems with the side jambs and head jambs. Check what the added width of the jamb area will do to the framing area.

When you are finished installing paneling boards, you can install the ceiling and floor and corner moldings. You can find suggestions for installing molding in Chapter 8.

Further suggestions

If using a large area of wide board paneling gives the room an aura of monotony, consider installing wainscoting. This is board paneling installed approximately halfway up the wall. The rest of the wall is papered or painted.

When wainscoting a room, install the wallpaper first so you will not have to worry about fitting the bottom of the wallpaper roll to the top of the wainscoting molding. Let the wallpaper come below the board paneling line and then install the boards so that the top edge laps over the wallpaper slightly. Then install the wainscoting molding over the ends of the board paneling, and the job is done.

Complete the blending work in the room by applying paint, varnish, or other wood coverings. You can make a room look slightly larger by using one flat paint and one glossy paint for wall covering. Use the flat paint for the wall surfaces and the glossy paint for window and door frames. You can paint the ceiling the same flat tones. If you want to give an impression of a longer wall, use a slightly different wall color from that of the ceiling.

Wide drapes give a feeling of expansiveness in a room, and you can achieve similar effects by using oval rather than square rugs on the hardwood floors. Throw rugs at doorways create a transitional effect as you move from one room to another. Paintings or wall decorations provide similar effects. You can also experiment with furniture arrangement to secure other desired effects.

Chapter 8

Finishing bookshelves, closets, and cabinets

When the rough carpentry in a room is completed, all of the framing is installed and the subflooring in place. Ceiling joists are nailed in, and all rough openings for doors and windows are completed.

The finish carpentry work includes installing all of the wall coverings, finish flooring, ceilings, and light fixtures. All doors and windows are hung and framed. Wall switches and receptacles or outlet boxes are installed, with cover plates in place.

You might still want to make additions to the room, such as a closet, a cabinet, or bookshelves. These extra items are very valuable to the comfort and utility of a home, and you can start on them as soon as the basic finish work is completed.

Before you do any work, measure the space carefully and buy your materials with equal care. When you were in the rough carpentry stages you had some leeway in that if you overbought on materials you could probably use them later. Excessive or surplus studs could always be used in the next room or project, as could drywall, ceiling tiles, or nails. However, now that you are near the completion of your work, you do not want to have your money tied up in materials that you have no use for, unless you plan to start a workshop or storage building later. While you will always have a few scraps of wood left over, the projects given here are relatively specific in terms of needs, so there is little chance of overbuying.

BOOKSHELVES

A bookshelf differs from a bookcase in that a bookcase is movable and generally freestanding. Bookshelves are not transportable and are usually attached to a wall or other permanent part of the room.

You can build a set of bookshelves to fill in an alcove, or you can cover a wall or all of the walls in a room. To build a set of shelves for an alcove, your work is simple.

Alcove shelves

You often find that when you install closets that do not extend the entire length of a wall, there will be a space left between the closet and the wall itself. Sometimes the presence of a window keeps you from building the closet to the full extent of the room. The resulting alcove between closet and wall can be an ideal place for a small set of bookshelves.

Starting with the alcove or recess in the wall, measure the distance to be covered by the shelves. You will need a vertical as well as a horizontal measurement. If the shelves are to extend from floor to ceiling, your vertical distance is probably 8 feet. The horizontal distance might be anywhere from 3 feet (or even less) to 6 or 7 feet.

As you begin to plan the work, calculate the number of shelves according to how you plan to use them. If the shelves are to hold books, how tall will the books be? Will the contents of the shelves be oversized books such as encyclopedias and special reference books, or will the books be typical novel size? Often such shelves are for students, and the books are frequently textbooks or reference books needed for classroom work.

Shelving should be no less than 1 inch thick. If you plan to shelve from floor to ceiling, determine how many shelves you can use most efficiently. When dealing with an 8-foot ceiling, you will be able to use only seven shelves maximum, with six being a more realistic figure. Remember that you do not put a shelf on the floor, or at the 8-foot mark. If the books to be shelved are more than 1 foot in height, you will not be able to have seven shelves.

As you work on shelves you will find it extremely difficult to install preconstructed shelves and get a good fit. If the shelves are already made, the depth of them will not permit you to slide or move the shelves into a standing position, unless they are built in that position. Shelves constructed flat on the floor cannot be stood into place because the depth of the shelves will not permit you to have ceiling clearance.

You can build the shelves in place rather easily if you follow these suggestions. Cut your vertical end boards to fit the space. They will probably be 8 feet long. Stand them one at a time into the space to see if there is a problem. If they fit well, lay each end board so that the inside surface is up. To choose the correct side of the boards, examine them visually for marring effects. Anything you do not want to be seen can be turned against the wall.

Locate the first shelf 4 inches off the floor. The space underneath will permit you to do the usual housecleaning to remove dust and other undesirable elements. If you have a small length of shelving board (2 or 3 inches), use this to mark the shelf locations.

Plan to put your heaviest and least-used books on the bottom shelves. These will often be your tallest books as well. Measure the height of the tallest books. If they are 1 foot tall and you allow an inch for finger space, plan to make the second shelf 13 inches higher than the first one.

Mark the location of the first shelf by measuring up 4 inches from each end and drawing a line across the board. The line represents the point where the bottom of the shelf will rest. Now stand the pattern length of shelving board so that the bottom of the pattern length is aligned perfectly with the line on the end board. Mark along the upper side of the pattern, and repeat on the other end board. You have now made the exact placement of the first board.

Calculate the height of the other boards by measuring the books you have to shelve and by estimating the books you plan to add to the shelves. Mark the height locations on the end boards as you did for the first shelf. Continue this process until you have reached the top of the end boards. If you learn that you will have several inches of unusable space at the top, go back and add an inch or so to the height of the first shelves. Erase the first marks and add the new ones.

How will you fasten the boards into place? One method is to drive nails or sink screws from the outside surface of the end boards into the ends of the shelving boards. This method does not always work with great satisfaction because the nails can miss the ends of the shelving board or split the ends of the boards. Also, if you decide to remove the shelf later you cannot do so without damaging it severely.

One nice solution to the problem is to install right-angle supports under the shelves. These supports cost very little and can be installed quickly. All you need to do is place the side of one angle against the end boards so that the top of the brace is even with the bottom of the shelving board line. Use two of these supports at the end of each board. When you are ready to install the shelving boards, screw upward into the bottom of the board. You might encounter difficulties with the lower shelves unless you install the shelves before you move the entire assembly into place.

One extremely easy and inexpensive method of installing shelves is to cut short lengths of end board stock and fasten these to the inside surface of the end boards. Cut the lengths so that they correspond exactly to the space between shelves.

Start with the bottom space under the first shelf. If the shelf is to be 4 inches off the floor, cut two 4-inch lengths and fasten these to the inside surface of the end board, so that the end of the support board matches the end of the end board. You can attach the support boards by using small finish nails (FIG. 8-1).

Measure and cut the next support boards, but do not install them yet. Stand the end boards into the alcove where they will be permanently

8-1 Nail the small support block to the bottom of the inside edge of the end pieces. When the shelves are in place they can rest on the support blocks.

located and spread them so that the outside surfaces are against the wall. Measure and cut the shelf board that will run from one outside board to the other.

When the shelf is cut, lay it across the support boards. You do not need to fasten it in any way. The weight of the books will hold the board firmly in place under normal usage. If you expect more than usual stress on the shelf, you can drive small finish nails through the top surface of the shelf board and into the end of the support board underneath (FIG. 8-2).

Measure, mark, and cut the next support boards and install these as you did the first ones. Then measure and cut the shelf that lies across the support boards. Lay the shelf across the support boards and fasten it with finish nails if you feel that it is necessary to do so. In this manner continue to the ceiling.

The support boards are now fastened to the end boards and the shelf boards to the support boards. Your entire book shelf unit is tied together well, except that the entire assembly is not tied to the wall.

Handle this final step by using the right-angle support braces. Attach one side of the brace (which is parallel to the floor) to the wall and the other side to the support boards. You can also attach the brace by turning it so that one side can be fastened to the bottom of the shelf and the other side to the wall. When you are fastening to the wall, be sure to attach the brace to a stud. Nails or screws in drywall will not hold under pressure.

8-2 Place the shelf on the support blocks and secure a good fit. The weight of the books hold the shelf in place, or use small finish nails to attach the shelf to the blocks.

Be sure that the support braces are installed so that they are as near the back of the shelves as possible. If they extend too far forward they can be unsightly and also get in the way of your books as you shelve them (FIG. 8-3). If you decide later to remove the shelves, you do not need to damage the boards or the wall significantly. Remove the screws holding the braces and, with a short crow bar or durable screwdriver, carefully pry the shelves free from the support boards. The bookshelves can be dismantled easily in a few minutes and installed elsewhere without having to be repaired.

When the assembly of the shelves is completed, you can stain or paint the shelves to suit the room decor. You will not be able to paint the back sides of the shelves or end boards, but these will not be visible and will not need decoration.

If you find that the shelves sag slightly under the weight of the books, cut lengths of support boards and install these at the midpoints of the shelves. One board will provide support for a 5- or 6-foot shelf.

Corner shelves

When you start to install shelves into a corner, you will find that the basic approach used in the alcove shelves works well, with slight modifications. The major difference is the use of triangular boards.

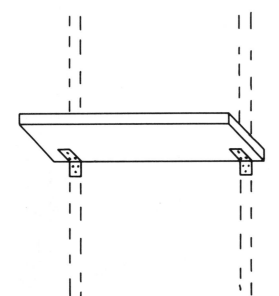

8-3 Attach shelves to the wall by first locating studding then using angled supports placed against the studs and the bottom of the shelf. Use screws for better holding power.

Select the corner distances you want for the front of the shelves. The first boards will be widest, and the one at the very corner will be the narrowest.

If the corner area you choose to shelve measures 4 feet diagonally across the front, your last board to install will be exactly 4 feet at the longest point. The first board to install will be a triangle, with the point facing into the corner.

The easiest way to lay out the shelving is to lay a series of boards on a flat work surface and pull them tightly together so that the ends are aligned. Use a straightedge to mark across the boards just to check that the ends are straight.

Assume that the ends of the board are to your right as you face the work. Measure diagonally across the boards to the outside edge of the board farthest from you. The distance should be 4 feet. If it is not, you will need another board or so. If you are using 10-foot boards, you will need only a few of these. When you have made the 4-foot diagonal line, saw the boards along the line. These sections will complete one shelf.

You will need a 4-inch-wide strip of wood to fasten to the base of the wall under the boards. Each strip should be 1-inch stock that is long enough to reach from the corner to the outside edge of the longest diagonal board (FIG. 8-4).

When these support boards are in place, start in the corner with the smallest board and install it so that the edges of the board are aligned perfectly with the wall and so that the edges rest on the support board underneath. Use small finish nails to attach the first board to the support boards.

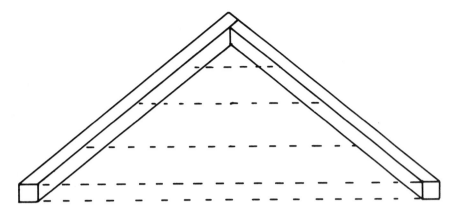

8-4 After the corner supports are in place, cut angled shelving boards and fit these into the corner. The first board should be a triangle, and then remainder of the board will follow the angle lines of this first board.

Next, install the second smallest board, which also should have its edges aligned with the walls and resting on the support boards. Continue in this manner until the final board is in place.

Cut the support boards that will separate the two shelves. If the shelves are to hold stereo components, allow for work room inside the shelf area. If the shelves are intended to hold VCR equipment, allow plenty of ventilation space.

When the support boards are cut, install them by nailing through the boards and into studs. Align another series of boards and mark and cut them as you did before. When this is done, install the boards as you did the first shelf (FIG. 8-5).

As you continue, you can make the space between shelves as narrow or wide as you want it to be. You should have some special use in mind

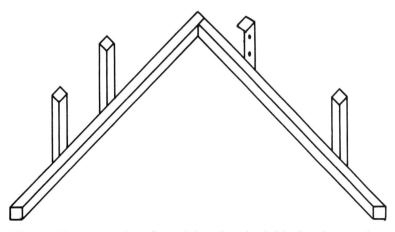

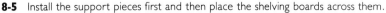

8-5 Install the support pieces first and then place the shelving boards across them.

for each shelf, because the depth will be too great for use for ordinary bookshelves. You may choose to use the space for flowers, photographs, or even a small aquarium.

Should you decide to remove the shelves later, you can do so easily by using a small pry tool and a small hammer. The boards will come out easily and the only damage to the corner will be the nail holes in the drywall and the discoloration where the support boards prevented light from reaching the wall space behind them.

Full-wall shelves

A full wall of bookshelves is a fine addition to any room. One wall of a small room will hold hundreds of books, phonograph recordings, and other materials that need to be stored so that they can be used readily and efficiently.

These full-wall bookshelves are very easy to install. If you hire a carpenter to install them, it will cost you hundreds of dollars. If you install them yourself, you will need seven 1-inch boards 8 or 10 inches wide and the length of the room. You can piece boards if you need to, but it is easier to use uninterrupted sections of boards.

For planning and instructional purposes, assume that the wall to be covered is 8 feet high and 16 feet long. Plan the full wall shelves basically as you did the recess or alcove shelves. Start at the floor level and measure to the top to be certain of the height of the wall in both corners. Then cut a board 8 feet long and stand it in one corner. Do the same for the other corner.

If you want a 4-inch space under the bottom shelf, cut five of these 4-inch-high support boards from bookshelf stock. Set three aside for the moment and nail two, one for each end board, to the bottom of the two boards you stood in the corners. The support pieces should be on the inside surface of the end boards. Place nails so that they will not show from the center of the room once the shelves are completed.

Now cut a full-length board that will reach all the way across the room. When it is ready, use finish nails to attach the end boards to the wall corners. They are now in place, flush against the corner wall. You can lay the board in place on the support boards at this time. Because of its length, the board will sag slightly. This sagging can be corrected by installing the remaining three support boards, one at the 4-foot mark, a second at the 8-foot mark, and the third at the 12-foot mark (FIG. 8-6).

The board is now supported at five points. If you choose to store extremely heavy materials on the shelves, add more support boards as needed.

Because you are using a full-wall bookshelf arrangement, you will want to make some of the shelves or portions of shelves high enough to hold phonograph recordings and oversized books or large photos. Place heavy items and those used less frequently on the bottom boards.

Start the second shelf by measuring, marking, and cutting the support boards for both ends. These can be as short as 6 inches and as tall as 2 feet or more, depending upon how you plan to use the shelves.

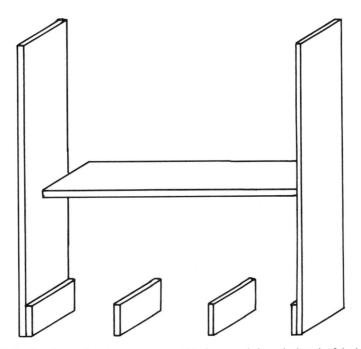

8-6 When installing wall shelves, use support blocks spaced along the length of the bottom shelf.

Place the first support board in position and attach it to the end board with small finish nails. Then lay the next shelf in place and attach it to the support boards by driving finish nails down through the shelf board and into the top edge of the support board. You can space other support boards as needed. A good idea is to place them, for perfect symmetry, directly over the first support boards. If you want a staggered appearance, space them at the halfway mark between the first support boards.

You can install shelves at alternate heights, even at midwall. You might want to store record albums but you don't need a full shelf. At the point where you feel you have adequate space, install a support board, which now serves as a divider as well. Against it install a shorter support board that is the proper height for the remainder of the shelf. Use short finish nails so there will not be nail points sticking through the two support boards.

There is an alternative to using full-length support boards. If you want to conserve boards, cut a 2-inch length of regular stock slightly shorter than the width of the book shelf boards. Be sure to cut it with the grain, or rip it, rather than crosscut it. If you are using 8-inch boards, cut the 2-inch strip 5 or 6 inches long. Fasten it at the desired height to the end board and then lay the next shelf board atop the 2-inch strip. You can use one or two small nails to fasten the shelf board to the strip just installed.

The major drawback to this approach is that the 2-inch strip will not support as much weight as the full-length support board. You can use the bracket angle supports if you want more support but do not want to use support boards.

When you reach the top of the shelves, you will have difficulty installing the final shelf because it will be flush with the top edge of the end boards. You can solve this problem easily by nailing the top shelf to the support board before you install it at the top of the shelves.

Nail the support board so that the nails enter the top of the final shelf to be installed. Do this on both ends of the shelf and in the middle, if you decide that additional support is necessary. Then, with the support boards in place, set the shelf in position. The ends should abut the end boards and the support boards should be flush against the end boards.

All you now have to do is drive small finish nails through the support board and into the end board. Keep the nails back from the edge so they will not be visible. Do this at both ends of the top shelf. When you are finished you are ready to add whatever trim is needed for the shelves.

Trimming bookshelves

In any type of finish carpentry you will have exposed edges of boards that might not be very attractive. This is particularly true of materials such as plywood. You might also choose to trim out the edges to hide the surfaces that clash slightly with the rest of the work.

You can use regular shelf stock to trim the shelves. Start where the support boards are installed. No matter how tightly you compress the two boards together, there will still be a visible crack. You can rip a board so that it will fit precisely over the two board edges.

Hold the uncut board or partial board tightly against the end board and the support board side of the shelves. Reach behind the new board and use a pencil to mark the cut line. Hold the pencil point against the support board so that the mark will be exact. You will have to interrupt the mark each time you come to a shelf, but you can use a straightedge to connect the marks.

Cut along the mark, and the finish strip will fit across the surface of the edges of the upright boards. Nail the strip in place using small finish nails. You can use a punch to drive the nails below the surface, then use plastic wood to fill in the nail holes so that no nails are visible.

You can also cut the finish strip slightly wider than the combined widths of the two upright boards. When you nail the strip in place it will provide a neat finishing touch to the upright portion of the shelves. You can also install finishing strips along the top and bottom shelves. Always have the top edges of the shelf and finishing strip flush so that books may be shelved or removed without difficulty.

When the shelves are completed, you can stain or paint them to fit the decor of the rest of the room. When you are using stains or varnishes, use a damp cloth to clean the surfaces thoroughly before you apply stain, and let the surfaces dry completely.

After you apply the first coat of stain, allow it to dry until it is faintly sticky or tacky and then rub it lightly with fine sandpaper. This will roughen the surface of the stain so that the next coat will adhere readily and provide a smooth, even, and uniformly colored surface. When you have rubbed or sanded the stain, let it dry completely, then add the second coat.

Use this same process with all shelves, cabinets, and other decorative or utilitarian additions you make to the finished look of a room. Be sure to let the coverings dry totally—at least 24 hours—before you add books to the shelves. If the stain is not fully dry it will stick to the bottom of the book covers and damage the books.

CLOSETS

If the closet space is already fully framed, just as you framed the walls, you can start to install the wall covers. Use the same materials that covered the walls in the room. The major difference is that walls all have inside corners and the closet, if it extends into the room or has a recess or alcove, will have an outside corner.

Covering walls

Install the paneling from the wall to the corner of the closet first, then install the paneling that runs parallel to the closet door so that the surface of the second panel will lap or cover the raw edge of the first paneling. Do this on both corners if the closet extends into the room.

If the closet is constructed so that the front walls are flush with the room walls, you can panel or install drywall as you covered the walls originally.

Doors

You will need to install doors for the closet, and here you have a number of options. You can choose traditional doors, double doors, sliding doors, or folding doors.

Sliding doors may be solid or have mirrors built in. They come both framed or frameless. They range in widths, unless you make a special order—from 47 up to 97 inches. The door package comes equipped with rollers and frames that can be installed easily. Complete instructions are provided for installation of the doors. You will need only a screwdriver and a hammer for most installation. You can expect to pay from $100 up to $250 for sliding or folding doors.

For sliding doors you must measure the space inside the finished door frame—not the rough opening—at the top, middle, and bottom of the opening. Measure all three times because of the possibility of irregularity in the framing. You can also put a level on the inside edges of the framing and check for trueness from both a horizontal and vertical direction. If you do not feel confident about the opening, use a square to check for trueness.

Some of these doors are designed to fit inside the door frames, and cannot be modified. With regular wooden doors you can plane off excessive thickness or saw off excessive height. Sliding doors, however, must be installed as they come from the dealer. That is why you must provide correct measurements and the openings must have square corners.

If your closet door openings are not correct, consider a traditional door, double doors, or wall-hung sliding doors. You can buy doors that have frames attached and these are large enough to cover the framed openings of most closets. These doors can be installed over the framing of the door as long as the opening is smaller than the door. Tracks can be installed quickly and easily by following manufacturer's instructions.

Folding doors can similarly be installed within the space of one hour, even if you have no previous experience in this type of work. All you need to do is read and follow the instructions provided with the doors. You can buy wall-hung doors for under $200 and folding doors for under $100.

CABINETS

You can buy ready-made cabinets that can be installed simply by holding the cabinets in place and tightening a few screws or driving a few nails. Or, with a little patience and planning, you can make your own cabinets.

Ready-made cabinets

Many people hire cabinetmakers to plan, design, make, and install cabinets for the kitchen, den, or bathroom. The cost of these special cabinets often runs as high as $3000 for one kitchen. If you choose to hire a cabinetmaker, check his previous work and get a list of clients you can call for verification of claims and promises.

Making your own cabinets

It is difficult to build really attractive and useful cabinets unless you have carpentry experience, but you can do it. You will find that plywood is one of the most common cabinet materials used in this country at present, which means that the cost of materials will not be unrealistically expensive. You will perhaps need or want to buy special glass doors or other similar products that you cannot make, but the majority of the material needed in cabinetmaking can be found in plywood and basic 1-inch boards.

The key to attractive plywood construction is the use of good stains to give the surface of the panels a good finish, as well as careful nailing or attaching so that no nails or screws show and no unsightly workmanship is apparent.

If you want to build cabinets to fit over the kitchen range, sink, and refrigerator, you can do so for less than $50 and your own time and expertise. First measure the distance you want to cover and by deciding whether you will need to fit the cabinet into a corner or create an angled cabinet that will fit in the corner and extend down both walls.

For illustration purposes, plan a cabinet that is 8 feet long along each wall, 20 inches high, and 1 foot deep. You will need plywood ranging from $5/8$ to 1 inch, depending upon your wants and needs. You will also need 4-inch boards, which are really closer to $3^1/2$ inches, small nails, glue (optional), a staple gun and staples (also optional), and a few additional elements such as laminated covering sheets.

Start by cutting a length of 4-inch board that is 7 feet, 7 inches long. Cut a similar length of board $2^1/2$ inches wide. Next cut four lengths of 3-inch-wide board 20 inches long. Lay the 7-foot 7-inch length face down on a good work surface and then place a 20-inch board so that the first board abuts it and the edges are even. Do the same at the other end of the 7-foot 7-inch board.

At the bottom of the assembly, lay out the length of $2^1/2$-inch board so that you now have a rectangle that is 8 feet long and 20 inches wide. This is the border of the front of the cabinet.

Cut from $2^1/2$-inch stock a length of $16^1/2$-inch material. This is the spacer for your first cabinet door. You can decide on the sizes of doors and door openings. You have a total of 96 inches of front space that will be devoted to doors and spacers, but 1 foot of this space will not be used in open space because the second half of the cabinet will abut it and utilize the final 12 inches of the first segment. You will then have 84 inches of door space.

There will be seven spacers, including the ones at each end of the cabinets, each $2^1/2$ inches wide. You will need 17.5 inches of front space for these spacers, leaving you with 66.5 inches of door space (FIG. 8-7).

You can divide the space as you need it. One suggestion is to allow for three door openings of 12 inches and two of $15^1/4$ inches. The two wider doors can come at each end and the more narrow ones can come in the middle portions for best symmetry. If your own preferences differ, use your judgment concerning widths and placement of doors.

Position the spacers at the locations you chose and then fasten the entire assembly using glue, nails, or corrugated fasteners. The finished product should be set aside for later use. Do not install the doors at this point.

Now construct a similar rectangle (do not include the door spacers) exactly the same size and shape. This will become the back frame of the cabinet.

When the frame is completed, cut the ends of the cabinet from the plywood stock or similar materials you selected for the project. These should be 11 inches wide and 20 inches high. Stand the two frames so that you can place the end inside the frames and fasten them as you did before. Do the same at the other end.

Installation

At this point you should have the framing of the entire cabinet section completed. Raise this assembled section so that it rests against the ceiling

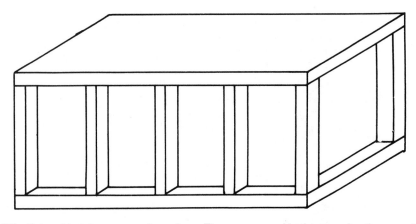

8-7 Cut and install spacers as shown here. These serve roughly the same function as wall frames. Doors can then be added, and the front of the cabinet is ready for finishing.

and wall. Drive nails through the back frame and into wall studs to hold the frame solidly in place. Do this at several points so there is no danger that the cabinet will fall.

It is a good idea to install the shelf supports before you install the entire cabinet. These supports can be 1-inch strips of molding attached to the ends and across the back framing at the point you want the shelves to be located. The first shelf should be located flush with the top edge of the bottom frame section. The second should be about the midpoint of the cabinet.

You can install the shelves at this time by cutting plywood sections that will fit inside the frame. Use glue or use small nails to fasten the shelves, but the weight of the items inside the cabinet will hold the shelves in place.

Repeat the first process to build the second framing, and when it is completed attach it to the wall as you did the first unit. Abut the second assembly to the first.

You can now cut the doors. The easiest way to attach doors is to use offset hinges that will permit the door to fit over the opening about 1 inch on all four sides. All you need to do now is attach the door to the hinges and the hinges to the frame of the cabinet and you have the basic cabinet completed.

You might want to stain the cabinet before you install the doors. When this is done, add the stained doors and the cabinet is ready to use. You can install molding around the outside of the cabinet where it joins the wall.

This is a very basic cabinet. If you want special design work, you can buy decorative doors. Your choice of hardware will also add greatly to the aesthetic appeal of the cabinets.

Chapter **9**

Carports and garages

*M*any home builders do not have the time or money to build the carport or garage at the time they build their houses. Other builders include details for a double car garage as part of the building project. To many people, the garage is a way of making the house appear much larger than it is. A 1500-square-foot house with a double garage can look like a 2500-square-foot house.

In many instances the garage is an excellent step in terms of economy. While the house is under construction you can get the concrete floor poured, the wall framing, the ceiling joists, rafters, and roof all completed for less money per square foot than the basic house required. Also, if you ever want to add living space to the house, the garage can be converted; all you need to do is the finish work to make the rooms comparable to the rest of the house.

Keep in mind that the cost of construction rises every time a new economic report is released from the government. In 1945 you could build a truly handsome house for $10,000. A house that cost $20,000 was a mansion in the eyes of many people. Today it is difficult to have a basic house built for less than $75,000−$100,000. The garage you add in 1990 for $15,000 will probably cost you twice that in 2010.

When you are constructing your carport or garage you probably want it to blend as well as possible with the house itself, even if the carport is not actually connected to the house. The best way to accomplish this end is to follow the same architectural pattern and use the same types or styles of materials. You can buy the same shade of paint and install the same basic roof style for starters. Other steps you can take to blend the two structures include installing ceiling tiles, paneling, drywall, and a disappearing attic stairway.

DISAPPEARING STAIRS

Disappearing stairs in the garage can open up some new storage space. Most disappearing stairways are made from either wood or aluminum. Wood stairways usually cost from $70−$165, while aluminum stairs sell for about $125.

As a rule, attic stairs are either double- or triple-fold. The number of steps can vary from eight or nine to ten or twelve. When you go to buy stairs, you need to know exactly how high your ceiling is. Many disappearing stairways are made for the typical 8-foot ceiling; if your ceiling height is different, ask the salesman for help in selecting the stairway.

Installation

To install the stairway, you will need a rough opening in the attic at least 25 × 55 inches. This means that you will need to cut a joist and install a header before you begin work on the installation of the stairway itself. Joists are, as a rule, 16 inches wide (on center), and at least one joist will have to be modified.

Cutting the joist can be difficult unless you stand on a step ladder and use a circular saw held vertically. *Note:* This can be a very dangerous step and should be attempted only with the greatest care. Wear protective eyeglasses and stand to the side of the work so that your body is clear of the blade.

If you decide to use a handsaw, use a level and square in order to mark the cut line. As you cut, follow the line carefully and do not allow the saw to deviate more than a fraction of an inch.

As you saw closer to the end of the marked line, the joist might sag and pinch the saw blade. If you are using any type of power saw this same problem can create kickback. You can eliminate much if not all of the binding by either lifting the joist with one hand while you saw with the other, or by using a shoring timber to raise the joist until the kerf or cut line opens to its normal expansion.

It is a good idea to shore up the joist on both sides of the cut line so that when the cut is completed the joist ends cannot fall abruptly. To shore up the joist, simply insert a board or larger timber under the joist and tap the end until the timber is wedged firmly under the joist. For the brief time that the timber will be needed, there is usually little reason to nail the timber in place, although you should do so if there is any danger that the timber will shift and allow the joist to drop (FIG. 9-1).

As soon as you have cut the joist at the first mark, move to the second location and mark the cut line. Your installation instructions should tell you exactly what length the rough opening should be. If the rough opening is 56^1/$_2$ inches, measure exactly that space and mark the cut line. Shore up the tail of the cut joist and make the second cut. The length of joist that you cut will fall, so be prepared to hold the joist section or have a helper who can keep it from crashing to the floor. If you must let it fall, clear the work area of all materials that can be damaged if hit by the falling joist section.

When the joist section is removed, you can build the rough opening for the stairs. Start by nailing in the header timbers, which are the same dimensions as those used for joists. Cut the first one exactly the length of the distance between the inside edges of the joists to the left and right of the cut joist. The distance between joists is probably 14.5 inches, so you

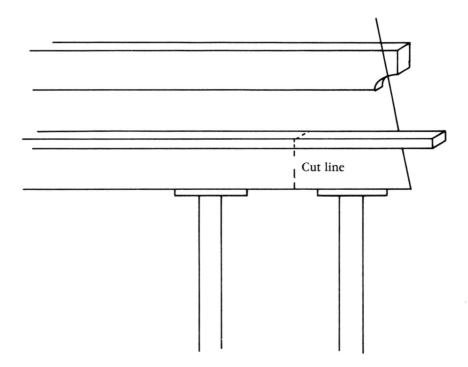

Cut line

9-1 Before cutting the joist, use support posts to support the ends until the work is completed.

will have 29 inches total space, plus the thickness of the cut joist—which will be about $1^1/2$ or $1^3/4$ inches—for a total of $30^3/4$ inches. Double-check the measurements to be sure (FIG. 9-2).

Cut the length of the header timber and install the piece so that the cut joist abuts it from the outside, while the header timber abuts from the side into the two adjacent joists. You can nail the header timber easier if you mark the location, then start two nails in each side and tap them until the points barely stick through the inside edge of the joists.

You can make the nail line by snapping a chalk line. Or use a long straightedge, held so that it is lined up exactly with the end of the cut joist, and mark across the top edge of the two adjacent joists. You can hold a short piece of header stock so that one edge is on the line and the other edge is toward the rough opening. Mark along the other edge of the header stock. Your nails then will be started between the two lines. Do this on both sides of the rough opening, and make the marks on the side of the joist away from the rough opening.

Repeat the process for the other end of the rough opening. Nail the header timbers to both joists and to the cut joist in the center. Be sure to nail both ends of the cut joist to the header timbers. You are now ready to install the remainder of the rough opening.

If your rough opening is to be 25 inches wide, and if the distance between the two adjacent joists is 31 inches, you will have 3 inches of

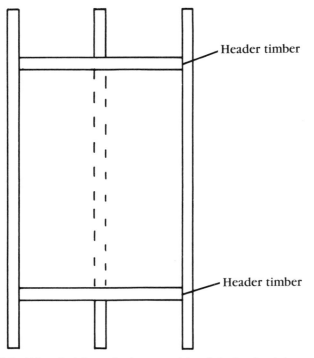

9-2   When the joist section is removed, install the header timbers.

space on each side of the rough opening. Fill this space by adding a joist timber the same length as the rough opening, less the thickness of the header stock. Simply measure from the inside edge of one header timber to the inside edge of the other header timber. Cut two lengths of joist stock equal to the distance measured.

Mark a point on the header timbers that is 3 inches from the nearest joist. Do this at both ends, then repeat on the other side of the rough opening. Cut the joist timbers and install them so that the inside edge of each timber is on the mark you just made.

Start the nails through the outside edge of the header timbers after marking the nail line. Hold the joist timbers and nail them to the header timbers. When this is done your rough opening is complete and you are ready to install the stairs.

On many stairway assemblies you will see metal hand rails that extend down to the third or fourth steps. Connected to the rail on each side is a metal support that extends horizontally. Either nail or use screws to fasten the arms to the sides of the rough opening. Follow the manufacturer's directions for the installation of the remainder of the stairway.

Fixing discrepancies

When the stairway is finished, try pulling it down. The stairs should unfold and the final length of the stringer rests against the concrete. Occasionally there will be a discrepancy between floor and steps, and you will need to fix this by adding to the stairway. You can do so by drilling holes in the final length of stairway and in a length of board the same dimensions as that of the stairway rails. Align the holes so that the necessary length of extension is left to reach the floor. Insert bolts into the holes and bolt the stairs and extension stock together. Check first to see that the stairs will continue to fold up neatly and disappear with the additional materials added (FIG. 9-3).

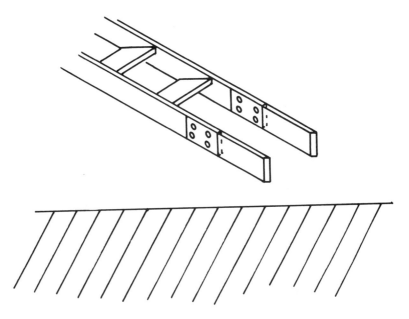

9-3   To extend disappearing stairways, either bolt additions to the existing stringers or bolt additions to an extender board. Then bolt the extender board to the inside of the stringers.

CONNECTING FREESTANDING CARPORTS

Often the carport is built apart from the house, but the distance separating them is not significant. You can make the carport blend with the rest of the house by carrying over some of the decor of the house into the garage or carport. One of the best ways to do this is with a covered walk or shelter connecting the house to the carport.

If the distance between house and carport is less than 16 feet, you can use a span of joist timbers to establish the basic connection. If the span is greater than 16 feet, in many areas you will need to install support posts. Many building codes stress that there can be no more than 16 feet of unsupported span in any structure, whether used as a dwelling or not, if the structural span is made of wood.

When you plan the covered walk or shelter, look for the best place to connect it to both house and carport. The most obvious place is above the door that is nearest the carport. A back door or side door will work well in most cases. Your first step is to establish a support base for the ends of the shelter. Do this by installing vertical timbers beside the doorway and a horizontal timber above the door, then connecting the timbers. Use 2-×-6 timbers for this purpose.

Install the vertical timbers on each side of the doorway, allowing plenty of access room. If your siding will permit, nail the vertical timbers to the siding. Be sure that the nails reach to the studding or other framing members.

If it is inconvenient to attach the vertical boards to the siding, you can use support timbers beside the door but based on the porch surface. You must have a solid foundation under the support timbers. Check your local building code for this and all other steps. If the support timbers must be set on the ground, you must dig and pour footings that reach below the freeze line.

For these support timbers you can use treated 4 × 4s. Keep the timbers plumb by installing bracing boards that are connected to the timbers and then to a stake several feet away.

Connect the tops of the support timbers by using spanning timbers. These can be 2 × 6, 2 × 8, or 2 × 10, depending upon the distance to be spanned and the local building regulations. Nail the spanning timbers to the outside edge of the support timbers. Do this on both sides of the walk between house and carport. One end of the spanning timbers should be connected to the house and the other end to the carport. When the timbers are in place, install the ridge board and the rafters.

You will want to use the same type of roof as that used on the house. See Chapter 6 for roofing instructions.

INTERIOR AND EXTERIOR WALL COVERINGS

You can cover the walls of garages and carports in a wide variety of ways. For interior walls these include plywood, paneling, drywall, boards, and wallpaper. For exterior walls choices include plywood, boards, shakes, vinyl siding, bricks, and stucco.

Try to match the exterior walls of the house when you cover the exterior garage or carport walls. Use the same style of bricks if you can find them, or match the exterior plywood carefully. If the exact style of plywood cannot be located, paint or stain the plywood you choose to match the house. If you choose one of the grooved or partitioned plywoods, you might have some difficulty in matching the style on your house, particularly if the house was paneled some years earlier.

Remember that plywood panels can be as attractive and economical as virtually any type of wall covering on the modern market. The insulation value of these wall coverings is also very efficient, and the panels are very easy to install. Also, one sheet or panel of plywood will cover 32 square feet of wall space; few building materials will cover quite as much with so little effort. When the paneling is in place, cover the cracks between panels with an exterior molding.

If your carport has a brick wall 2 or 3 feet high, you might want to match the carport to the wood siding on the house. Leave the brick partial wall, install the framing and studwork for the siding wall, and install a drip board at the very top of the brick wall so the water from the side of the house will be diverted over the edge of the bricks.

On the interior walls of the garage or carport try to create a logical stylistic lead-in from the house to the garage. If the house is Early American inside, use pine or knotty pine boards for wall covering. Use the same type of boards on the ceiling as well as the walls. Install antique-looking switch plate covers for the garage lights, and use a hanging fixture on a very short chain for the ceiling. Use a fixture that is not dressy but that follows the trend of the interior of the house.

While the house floors might be carpeted or composed of hardwood, in the garage suggest the same decor by using exterior throw rugs or area carpets that blend well with the interior flooring. If you cannot use any hardwood flooring, you can use simulated hardwood stains for the door framing of the outside door.

Door knobs, hinges, and other hardware for doors can be good lead-ins. Do not attempt to blend the two areas that are obviously contrived. Let the lead-ins be subtle and faint rather than glaring. Decorations on the walls can also tie together interior and exterior rooms. Shelves can be painted or stained appropriately, and if the carport walls are brick and the interior walls are boarded, you can use wide molding of the style used inside for the blending effect.

INTERIOR AND EXTERIOR MOLDING

One of the very best ways to finish a room is with molding that links floors and walls, walls and ceilings, and in corners and beside windows and doors. This molding serves several purposes. One of the most important is to conceal cracks where drywall, paneling, or boards do not fit perfectly against window frames, floors, or ceilings.

Use small finish nails to install the molding. Use only as many nails as necessary. Too many nails are unsightly, and nails that are too large can split the molding.

Some decorators like to use a baseboard to blend walls and floor, and then use floor molding along the bottom edge of the baseboard. In many modern houses both baseboard and molding are omitted, but if you omit the molding you will need to have near-perfect workmanship where the walls meet the floor.

The only difficult aspect of installing molding is that of cutting the molding properly. You will need a miter box for the best cuts. If you do not have a miter box you can buy one fairly inexpensively, or you can make one.

Making and using a miter box

To make a miter box you will need three lengths of 1-x-4 board—each a foot to 15 inches long, some nails, a square, a hammer, and a saw. Saw all ends of the boards square and make all three the same length. Lay one board flat and stand the other two so that their bottom edges are against the work surface and snugly against the flat board.

Nail the two side boards to the bottom board so that you have formed a square U. Use the square to mark a straight cut across the two upright board edges. Move 4 inches away and measure the exact distance from the outside edges across the miter box. Use the square to lay off a square with the same dimensions as the distance you just measured. Then use the square to mark diagonally from both corners of the square.

Saw along the first marks you made through the upright boards and down to the bottom board. Do the same with the diagonal cuts in both directions on the corners of the square you laid off (FIG. 9-4).

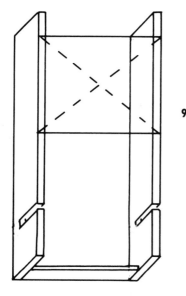

9-4 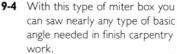 With this type of miter box you can saw nearly any type of basic angle needed in finish carpentry work.

When you use the miter box, lay the stock to be sawed inside the box and hold it snugly against one of the side boards. Insert the saw into the cuts you made and saw through the molding stock. Using the miter box you can make most of the necessary cuts you will need when installing molding. You will, in fact, be able to use it to cut any angle you encounter in finish carpentry work.

Installing molding

As you are installing molding around doors or windows, position the molding strips so that both ends extend slightly past the corners. Mark the corner points on the molding. Then lay the molding in the miter box and make the necessary left-hand and right-hand cuts.

When you install the molding, the points left from the cut will stick up slightly higher than the window frame. The top molding will be cut so that it will fit neatly into the space left.

Hold the top molding in place and mark where the corners of the windows strike the molding. Use the miter box again to make the cuts, then install the molding. With a little practice you can make all of the miter cuts without hesitation and fear of mistakes.

One of the easiest ways to install ceiling or floor molding (as well as several other types) is to lay a length of molding on a work surface and then hold a second piece at right angles to the first and abut the two. Mark the line where the second crosses the first. Cut along the line and then install the molding sections. The two pieces should fit well together.

If you cannot cut a molding fit that is satisfactory, make the closest possible cut and then fill in any spaces with plastic wood. Sand the filled areas so they blend well with the molding.

Chapter **10**

Porches and decks

A somewhat sad note in modern American building is that porches are often omitted from new houses. A stoop or tiny porch to serve only as a rain shelter while entering the house is substituted instead. A happy note, however, is that many houses are adding decks for leisure activities.

The argument against porches is that it costs as much to roof and floor an area that is used only a few minutes each week as it does to roof and floor a room that will be used constantly by the entire family.

An argument in favor of porches is that they provide a great deal of shade for one side of the house and thus will help to reduce the cost of air conditioning. Porches also protect large expanses of wall space from dew and rain and help to protect the paint from excessive exposure to the sun. Yet another good point about porches is that if the need arises for another room, it is usually a simple matter to close in the porch and make an extra bedroom or family room quickly and inexpensively (FIG. 10-1).

Whatever the advantages and disadvantages, if you have a deck framed or a porch ready to finish, the final work is simple, easy, and rewarding. All of the basic framing and rough carpentry had been done. Your job is to apply the final touches.

FLOORING

Assuming that subflooring is in place, you need to install the final flooring. In many porches no subflooring is used; you use instead the thicker decking boards and omit the cost and labor of installing subflooring.

The decking boards are referred to as five-quarter boards, which means that they are 1¹/₄ inch thick—nominally. Nominal size refers to the size before they are finished and dressed; after the finishing they are only slightly more than an inch thick. This thickness is not sufficient to hold the weight of a fairly large adult without some giving. To eliminate this uncomfortable aspect of a porch or deck, there must be adequate support.

10-1   A traditional porch can be closed in and converted to a sun room very easily. Such a room offers a warm sitting area in cold months and also helps contain heat.

Joist spacing

A wide board stretched over a span of 10 or 12 feet sags in the center of the span and will not support much weight. If the span is reduced to 2 feet or less, the same board will support several times its previous capacity.

This principle holds true with decking boards. If joists are spaced too far apart, the board weakens in the middle of the span. One of the major arguments in favor of subflooring is that the plywood or other boards that run at right angles to the finished flooring give considerable support, so there is no sag or give on the porch or deck.

Assume that your house runs east and west and that the joists for the porch run north and south. If the joists are spaced 16 inches on center and no subflooring is used, you will have slightly more than 14 inches of unsupported board between joists. If a heavy person steps so that his foot is supported by only one board, there will be a certain amount of give.

You can correct the problem by installing horizontal Z bridging, but this is time-consuming and fairly expensive. To construct this type of bridging, cut two 1-×-4 boards that will fit snugly between joists and install these 4 feet apart by driving nails through the opposite side of the joists and into the ends of the bridging boards. You can use larger lumber if it is available. A 2 × 6 will work much better but costs much more (FIG. 10-2).

10-2 This type of support system will strengthen long unsupported sections of a porch or deck.

When these two boards or timbers are installed, measure diagonally across the rectangle and cut a timber that will fit into the corners of the rectangle. Install the timber by holding it so that the top edge is flush with the top edge of the first lumber installed in the bridging. Drive nails through the corner of the first bridging, into the new pieces, and into the joist.

If you wish, you can install the diagonal sections first and nail them directly to the joist. Then add the right-angle sections. The completed support resembles a Z slightly.

A simpler way to add support is to nail up the cross bridging and then cut and install one length of 1 × 4 or 2 × 4 (or heavier timbers if they are available) down the center of the span between joists. This type of horizontal I support works well.

If the space between joists is greater than 16 inches on center, you might need to add a joist between the existing joists. Your local building code might specify that joists as large as 4 × 8 inches can be spaced as far as 3 feet apart. Do not settle for such spacing unless the decking boards are much thicker than the five-quarter boards. What matters most is not the size of the joists but the span between the joists. If the joists were 10 × 10s spaced 5 feet apart, the porch or deck would not be safe or serviceable unless the decking lumber is at least 2 inches thick, and preferably thicker. Very few building codes will permit such joist spacing.

Installing decking boards

When the joists are properly spaced and the framing is done to your satisfaction, you can start installing the decking boards. If you are working with a subfloored porch, you might want to use tongue-and-groove lumber. Start at one side of the porch and turn the groove away from you as you work so that the tongue side is near you. If you turn the boards the opposite way there will be no nailing space (unless you don't mind if the nail heads show). You can use nails through the top of the first board, but for subsequent boards nail only through the tongue.

With the first board in place, shove the next board against it so that the groove of the second board slips snugly over the tongue of the first. If the two boards will not fit together well, check the tongue and groove to see if either has been damaged. Make the necessary corrections.

If the boards will still not fit together but there are no signs of physical damage, align the boards and place a small block of wood against the tongue of the second board. Tap the block of wood with a hammer. This added force is usually sufficient to seat the tongue into the groove. Then nail through the tongue. Space nails 2 or 3 feet apart along the length of the boards.

Continue across the entire expanse of the floor or deck. when you are ready to install the final board, you can trim, plane, or saw the tongue section off for a neater appearance.

If you are using decking boards or timbers without tongue or groove construction, simply position the boards in place and nail them to the joists. You might want to leave a tiny space between the boards so that moisture may quickly flow from the surfaces. If you leave the boards tightly pressed together, standing water can cause warping, swelling, and decay as well as attract insects to the moisture-laden wood.

If there is still a problem of sagging after the flooring is completed you can work under the floor and install the extra support. When the floor is to your satisfaction, add the railing or bannisters.

ADDING RAILINGS

If your structure is an elevated deck, railing is a necessity. The size of the railing timber is optional, but use nothing less than 2 × 4s or 2 × 6s. Even 4 × 4s can be considered. This railing will be attached to the joists or header timbers and run parallel to the header timber. It will extend 3 or 4 feet higher than the floor. If you do not want a perfectly vertical railing structure, saw the section of the railing post that will fit against the header or joist at a slight angle, so when it is installed it will lean slightly outward (FIG. 10-3).

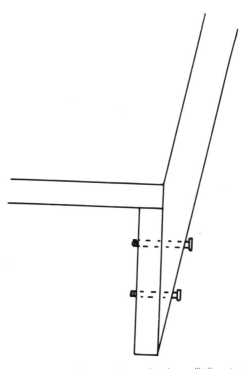

10-3 By sawing the supports for rails at a slight angle, they will tilt outward for an attractive appearance.

To get the proper angle, lay the corner post on a flat surface and mark a line from the bottom edge diagonally to a point on the inside edge that is at least the height of the joists. If the joists are 2 × 10s, the diagonal line should be 10 inches long. Start by making the line 1 inch from the corner. After you have sawed the corner off, hold the railing post in place to see if the angle is satisfactory. If it is, saw all other railing posts in the same manner.

You will place railing posts about every 6 feet. Install the railing post with either very large nails or nuts and bolts. If you use nails, there is a chance that they will split the railing posts unless the posts are thicker than 2 × 4. To keep the railing posts from splitting, use a drill with a small bit—one with slightly smaller diameter than that of the nails—and drill a hole where the nails will be installed. When you drive the nails the pre-drilled hole will allow entry without splitting (FIG. 10-4).

10-4   For a straight railing system, cut the posts straight and use either large nails or bolts to hold them in place. With large nails, drill a hole in the wood to keep the nail from splitting it.

If the points of the nails stick through the other side of the joists, use a hacksaw blade to cut off the points close to where they emerge from the wood. After you have cut off the points, use a hammer to bend the short protrusions down and drive them into the wood at an angle. This will prevent the nails from pulling out when there is pressure against the railing.

Some people want to cut their own lumber from their own trees. Many building codes require that, unless you use graded lumber, the lumber will have to be tested for moisture content. The moisture level in the lumber cannot exceed 19 percent. What is especially ironic about the rulings is that your own lumber can be rejected because it has a moisture content of 20 percent, while the graded lumber you buy from the dealer can have moisture contents up to 35 percent. Because it is graded the dealer's lumber is acceptable, while your own, which may be far superior, will be rejected.

The problem with a high moisture content, in addition to the fact that the wood will attract insects such as termites, is that the wood will warp and shrink as it dries. When you drive a nail, the force of the nail entering the wood separates the wood fibers, which in turn push vigorously against the nail shank and hold it in place. When moisture-laden wood dries and shrinks, the fibers holding the nail in place also shrink and the nail loosens.

This is the reason it was suggested earlier that you bend the end of the nail over slightly and hammer it into the wood. Then, even when the fibers of the wood shrink the nail cannot pull loose. It might lose some of its holding power but it will not work its way out of the wood.

To be completely safe you can use bolts. These are more expensive than nails but have their own built-in protection against the shrinking of wood.

To install bolts, first drill a hole completely through the railing timber and also through the joist. Push the bolt all the way through both timbers after first slipping a washer over the end of the bolt and sliding it against the bolt head. When the other end of the bolt emerges from the inside of the joist, put on another washer and then add the nut. Tighten the nut until the pressure begins to crush the wood fibers slightly.

Later check the tightness of the nuts. If the wood has shrunk slightly, tighten the nuts more. Check the bolts every month or so until the wood has cured completely.

After you have installed all of the railing posts or uprights, add the railings themselves. You put these up either with nails or with bolts. You can install the railing boards either on the inside or outside of the posts. The advantage to installing the boards on the inside is that when pressure is applied by people leaning or resting against them, the boards cannot fall even if the nails are loosened because they are supported by the posts (FIGS. 10-5 to 10-7).

You might eventually decide to close in the deck or porch by adding a roof and screen wire. If the floor boards have spaces between them, the area will be susceptible to insects. Some builders leave spaces an inch or even wider. Such spaces will permit high heels to sink between them and

10-5   If the deck is to have more than one level, this basic pattern can be used to install a short set of steps.

can be dangerous. Even small rodents and some snakes and lizards can crawl through the openings. Spaces are should be no more than $1/2$ inch. The rule of thumb is that gaps between boards should be the thickness of a quarter, if the reason for spacing is to allow water to drain and air to dry the spaces between boards.

10-6 Join railing to posts by abutting 2 × 4s to the posts and toenailing them.

SEALING WOOD

Any wood that is exposed to water frequently and for prolonged periods of time should be treated against moisture damage. All deck flooring lumber should be professionally treated; if it is not you should treat it quickly after installation.

One of the products now used commercially for wood preservation is borate, which is reportedly totally safe for people and animals but

10-7 Use slanted 2 × 4s for railing for steps.

deters insects and fungus. The product is said to have about the same toxicity as table salt.

Another product that has received considerable acclaim is the water seal line. There is one product that can be used on concrete and a variety of other surfaces and another product made and marketed by the same firm that is designed specifically for wood. Those who have used the wood sealer seem to be virtually unanimous in praising its abilities to protect wood from penetration by water and subsequent rot and insect damage.

Most water seal products can be applied with a brush, just as you would apply paint. When you are applying water seal, dip the brush so that half of the bristle length is submerged. Hold the brush slightly above the fluid level and allow it to drip for a moment or two. With the brush well loaded, brush it on the wood surface in the direction of the grain. Coat the wood well, particularly where there are cracks or the start of cracks. Many types of wood are prone to cracking as they cure, and these cracks are perfect places for insects or fungus growth to start. Do not skimp on the sealer. Get a full and thorough covering on the wood, and later, if necessary, apply the sealer again. Follow the directions on the label. It is much cheaper to protect and save the wood than it is to replace it later.

SCREENING IN PORCHES

Although porches are wonderful, you might decide that it would be so much nicer if you weren't bothered by insects and other pests. If you choose to screen in the area, you will find that the work is simple and it moves very quickly. You will also find that the porch is twice as enjoyable as before.

If the porch extends across the entire front, side, or back of the house, measure the length of the porch on all sides and divide the number of feet by four. Screen wire comes in 4-foot widths as well as in other sizes, so you can decide how many widths you need.

Calculating amount of screen needed

Wire also comes in rolls, and you will need to add all of the widths and then multiply by eight, which is the number of feet in the height of the porch. If you have a porch that is 10 feet wide and 40 feet long, you have a total of 60 feet. That means that you will need 15 widths of screen wire. If the ceiling is 8 feet high, you will need 120 feet of wire (15×8) for the entire porch.

The so-called "screen wire" is actually a vinyl mesh. Wire as such has not been used in years. You might be able to find some dealers who still have the actual wire in stock, but more likely you will find one of the newer synthetic products. Synthetic wire is much easier to work with and to maintain. It does not rust, and you can cut it with a pair of scissors very easily. You will not have to endure the painful scratches you get from real wire, and the synthetic wire will not rust and discolor the side of your house after a rain.

Placing porch posts

After you have purchased your wire, the next step is to set up the porch posts. You should use 4-×-4 posts if your budget will permit. If not, use a 4×4 alternating with a 2×4. At the very least, use a 4×4 or a doubled 2

× 4 on both sides of the door (if there is one), and a 4 × 4 at each corner and where the porch joins the house.

Set these posts so that they are 4 feet or 48 inches on center. When you set up your corner posts, measure from the outside edge of the post to a point 48 inches across the porch. Set up the second post so that the exact center of the post comes at the 48-inch mark. The posts must be the same distance apart at the top and bottom. Measure at the center to make sure the posts don't curve (FIG. 10-8). Use a level to determine that the posts are plumb. If you do not maintain proper sparing, the screen wire might not reach all the way across the opening.

Put up all the posts, using the same process, then install the wire. Or, begin installing wire after the first two posts are in place. In either case you should measure at several points between the top and bottom of posts to see the they do not bend and cause the alignment to vary.

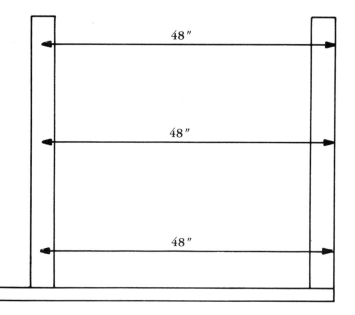

10-8 Be certain that posts are the same distance apart at the top and bottom. Measure the center to be sure that the posts do not curve. If a proper spacing is not maintained, the screen wire might not reach all the way across the opening.

Installing the wire

Start at the top of a post and be sure that you get off to a square beginning. If not, the wire will run off one side or the other as you move down. Cut an 8-foot section of wire, perhaps allowing a little extra in case of a problem, and staple the wire in place at one top corner. Then pull the wire tight across the top and staple the second top corner. If all still looks well, staple at several points across the top. You should have a staple at no less than 12-inch intervals.

Now move to the bottom of the posts and pull the wire tight. Do not put enough tension on it to stretch it out of shape, but enough that there are no sags in it. If it fits well at the bottom, staple the first corner. Then pull the second corner tight and staple it, and staple across the bottom.

You are now ready to staple up the posts on both sides. If the wire tends to sag or hang loose at any point, pull it gently tight and staple while you hold it.

Measure the distance between the two posts at the center and mark and cut a bridging timber to be installed. Drive nails from the outside edge of the erect 2 × 4 into the end of the bridging timber. On the 4 × 4 side you will need to toenail or angle-nail the timber in place (FIG. 10-9).

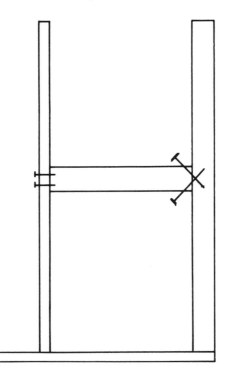

10-9 Install bridging by nailing through the support posts and into the bridging. If support posts are too thick, toenail the bridging in place.

The purpose of this timber is to keep the posts from bending as they continue to cure. You can install two timbers, one at the one-third point and the second at the two-thirds point, to give you better support.

After you have installed the first bridging supports, measure the distance from the floor to the bottom of the timber and then cut a length of board that is that exact length. Later you can stand the board against the post and lay the bridging timber on top of it. This will give you a good nailing support and you will not have to worry about correct measurements.

You can nail one end through the post, but because of the bridging timber just installed, you will have to toenail the other end all the way across the porch. The other option is to stagger the heights of the various timbers so that you can nail through the post and into the ends of all the timbers.

One suggestion for staggered heights is to let the top edge of one timber come to the bottom edge of the one beside it. Another is to use one timber in one area and two in the next. This way you will have a symmetrical design that is attractive as well as efficient.

If the ends of the porch are 10 feet long, you will have one 2-foot and two 8-foot panels on each end. Cut the length of wire down the exact center and install half of it on one end and the other half on the other end. To get an exact cut, lay the wire on a work surface and place a straightedge down the center so that one edge of the straightedge is on the exact midpoint. Use a knife with a strong blade or scissors to cut along the edge. You can also mark the midpoint with a chalk line or colored pencil and cut along the line.

Adding trim strips

When all the posts are in place and all the wire is installed, you will probably want to add trim strips. These are very thin strips or laths of wood that are usually about 2 inches wide. You cut and fit these to the outside edges of the posts and then nail them in over the wire. These lath strips will cover the unattractive staples and add support to the wire fasteners.

You might want to consider painting the posts and bridging timbers before you install the wire. No matter how careful you are, you will probably get some paint on the wire, and it is very difficult to remove all of the paint stains. Also, paint the lath strips before installing them. Use small finish nails and you will not have to be concerned about the unpainted nail heads.

FINISHING UP

At this point your work is done except for the final touches: painting or staining and other blending efforts. Paint the porch posts and molding the same color as the house. If the house already has more than one color, let the porch blend or match with one of the dominant colors. If the house is white and the shutters are brown, you might want to paint the porch posts and molding the same shade of brown as the shutters.

If windows are trimmed in a different color, match the porch molding and posts with the window trim color. If there is a deck on another side of the house that is stained rather than painted, you might want to stain the porch posts and molding to match the deck stains. If you use redwood for decking, you could use the same wood for the porch posts.

Many parts of the house can be blended. At times you might wish to add shutters, install brick steps, hang a light fixture, choose a different shade or style of shingle, add wrought iron railing, install heavy wooden bannisters or rails, change the color of the door and window framing, or in dozens of other ways make the house units lead to one harmonious theme.

There are no specific directions as to how this blending should be done; every individual will have specific ideas. It is your house and you have the right to please yourself and family. However, keep in mind that if you ever decide to sell the house, the decorations should be generally appealing.

Concern yourself mainly with any decorative steps that cannot be altered if your tastes change. Paint and stains can be easily changed, as can door and window framing, the doors themselves, light fixtures, shutters, wallcoverings, and similar elements of the house. It is more difficult, but not impossible, to change window styles. It is very difficult to make changes to parts of the house such as the roof, concrete work, and brick masonry.

When you are finished with your work, you will have every reason to be proud of the efforts, the ingenuity, and the taste with which you accomplished your goals. Your sense of pride and accomplishment will repay you many times over for the hard work, the sweat, the frustrations, and the exhaustion your efforts cost you. You will also have proven to yourself that, although you started as an amateur, you are capable of professional workmanship.

Glossary

adhesive   glue-like substance used for such building purposes as adhering wallpaper to prepared wall surfaces.

aggregate   any material, usually sand or small pebbles, that is mixed with the cement and water.

angle iron   A length of metal that will support the weight of the building above it. The angle iron is installed in the mortar joint across the tops of windows.

angle-nail   nailing two pieces of wood so that the nail point is driven through one side and out the bottom or top of the board rather than out the opposite side.

apron   a milled unit of lumber usually from 2 to 3 inches wide that fits into the window sash area to prevent rain or wind from entering the structure.

asbestos shingles   fireproof roofing shingles that are shaped and installed similar to wood shingles.

bannister   the railing along the stairway.

baseboard   the board running along the bottom of a wall and at a right angle to the floor.

batter boards   boards used in masonry to establish a framework to mark the corners of foundation that is being excavated.

binders   the ingredients in a mortar or plaster mix that hold the combination together.

blind stops   the pieces of lumber in a window assembly that are installed between the side jambs and the outside side casings.

block line   a line or slender cord used in masonry construction to provide the mason with a guide for the brick or block positions.

bonding   a tying of blocks or bricks that prevents the courses of masonry from sagging or falling.

borate   a substance used for wood preservation that deters insect and fungus growth damage.

bottom rail   the lower rail of a window assembly that joins the lower side stiles.

brick facing   the process of using bricks to complete or cover a traditional wall.

bridging   metal or wooden supports installed between joists.

C clamp   a C-shaped device with an adjustable screw used to clasp materials during construction.

ceiling, suspended   a ceiling that is built lower than the original room height.

cementicity   in masonry the ability of one matter to adhere or cling to another.

clapboard   the exterior wall covering consisting of boards that are usually 4 or 5 inches wide and 1 inch thick or less.

compound, ceiling   a caulk-like material applied to ceilings to give an appearance of plaster.

corner boards   two boards located on the corner of the wall. Usually one is 4 inches wide and the other 5 inches wide.

corner joints   connectors, usually shaped at right angles, located at the corners of a roof to join guttering troughs.

couplings   connectors that join straight lengths of guttering troughs of a roof.

cripple studs   short studs used below and above windows and occasionally above doors.

cross bridging   boards used as supports that are installed at right angles to parallel joists.

cut nails   nails used in tongue-and-groove carpentry that are manufactured with squared or flat sides.

dado cut   a rectangular groove cut in the side of one board for the purpose of fitting another board to it and at right angles.

elbows   part of a guttering system that can be turned to provide the exact angle needed to connect the guttering trough to the downspouts.

ells   the L shaped connections used in guttering systems.

end caps   the units with closed ends that are installed at the end of each guttering trough.

feathering   a process in which a joint-filling compound is spread so thin that the compound becomes only a smear.

ferrell   a long, narrow metal unit used to hold the spike that supports a guttering trough.

float   a device in masonry that is used to slide over the surface of the mortar or plaster and to fill in any holes, cracks, or thin spots in the work surface.

footings   trenches dug below the frost line that are filled with concrete and become the foundation of a house or building.

furring strips   thin lengths of wood that reach from ceiling to floor and have horizontal strips connecting them.

grout   a thin mixture of mortar used to fill in between tiles or as chinking.

gypsum   a substance that, covered with paper, forms wide sheets that are used as wall coverings.

hawk   a special mason's trowel that is a square sheet of thin metal or flat wood with a central handle.

head jamb   a board that forms part of the interior surface of the doorway and is located at the top of the door.

head joints   the vertical joints of all masonry blocks or bricks that match in every other course.

header   a structural member placed horizontally over doors, windows, or other framed openings to carry the weight of the opening.

jamb, side a board that forms part of the interior surface of the doorway and is located at the sides of the door.

joists, ceiling wide boards that form the framework for the ceiling of a room.

joists, floor wide supporting boards that form the framework for the floor of a room.

lintel a length of angle iron or similar metal installed over windows or doors, with masonry construction supporting the weight of the building above it.

lower stile the strips of wood, metal, or vinyl of the lower half of a double window, located on each side of the window.

meeting rails the lower strip of wood, metal, or vinyl of the upper half of a double window.

metal flashing a strip of metal installed in the valleys between the normal roof slope and the gable or similar construction areas.

mortar bed joints where the masonry block or brick joins the mortar.

muntin a vertical window strip.

parget a thin mortar or stucco.

pargeting spreading thin mortar mixture over a masonry wall for the purpose of decorating and waterproofing the wall.

plasticity in masonry, the workable quality of stucco or mortar.

rafters timbers in construction that reach from the ridge board to the top plate.

ridge board lengths of boards that reach the entire peak of the roof and the point at which every roof rafter connects.

roof cap a length of metal that is shaped to fit over the roof peak and lap the edges of the roofing metal sheets that end at the peak.

roof sheathing any covering of the rafters to which roofing shingles are nailed.

sisal hemp a fiber used for decorative purposes in wall coverings.

stiles the vertical boundaries of a window.

story pole a length of 2 × 4 or other timber used by masons to measure the height of each course of blocks or bricks.

striker plate the latch receptacle of a door.

stucco a mortar mixture used as a decorative wall covering.

studs the vertical timbers between top plate and sole plate that form the wall frame.

subflooring the first layer of flooring placed on top of floor joists.

template, lock a silhouette of a lock used as pattern for installing a lock.

threshold a length of metal or wood covering the bottom of a door opening.

toenailing angle nailing.

top cap the timber that is fitted along the top of the top plate.

top plate the timber that spaces and supports the top of the studs.

top rail the horizontal frame of the window.

upper meeting rail the top of the bottom half of a double window.

wainscoting a decorative technique in which paneling is used on the lower portion of a wall, and wallpaper or similar material used above the paneling.

Index

Other Bestsellers of Related Interest

KITCHEN REMODELING—A Do-It-Yourselfer's Guide—*Paul Bianchina*

"*. . . offers all the know-how you need to remodel a kitchen economically and attractively.*"

—*Country Accents*

Create a kitchen that meets the demands of your lifestyle. With this guide you can attractively and economically remodel your kitchen yourself. All the know-how you need is supplied in this complete step-by-step reference, from planning and measuring to installation and finishing. 208 pages, 187 illustrations. **Book No. 3011, $14.95 paperback only**

KITCHEN AND BATHROOM CABINETS
—*Percy W. Blandford*

Kitchen and Bathroom Cabinets is a collection of wooden cabinet projects that will help you organize and modernize your kitchen and bathroom and make them more attractive at the same time. Clear step-by-step instructions and detailed drawings enable you to build wall and floor cabinets and counters, corner cupboards, island units, built-in tables, worktables, breakfast bars, vanities, and more. 300 pages, 195 illustrations. **Book No. 3244, $16.95 paperback, $26.95 hardcover**

THE WOODTURNER'S BIBLE—3rd Edition
—*Percy W. Blandford*

Long considered the most comprehensive guide available on woodturning techniques, this book is an authoritative reference to every aspect of the craft, from choosing a lathe to performing advanced turning techniques. Blandford covers every kind of lathe and turning tool available, then gives step-by-step instruction on tool handling techniques and lathe applications. This expanded edition offers added coverage of turning wood patterns for metal-castings and tips on using high-speed steel tools. 272 pages, 210 illustrations. **Book No. 3404, $16.95 paperback, $26.95 hardcover**

24 WOODTURNING PROJECTS
—*Percy W. Blandford*

Here are 24 ways to put your lathe and your woodturning skills to work creating beautiful and useful objects from wood. You'll find projects ranging from the simple to the more advanced, including such items as candlesticks, bowls, floor lamps, canes, a round table, boxes and lids, vases, jewelry, and more. Each project provides a materials list and advice on wood choice, step-by-step instructions, detailed drawings, and a picture of the finished piece. 140 pages, 95 illustrations. **Book No. 3334, $9.95 paperback, $18.95 hardcover**

DOORS, WINDOWS & SKYLIGHTS
—**2nd Edition**—*Dan Ramsey*

There's no end to the money you can save and to the value and comfort you can add to your home with the practical ideas found in this well-illustrated guide. Ramsey gives step-by-step installation instructions for all kinds of attractive, economical interior and exterior doors, windows, and skylights. This edition features information on designing solariums and greenhouses, increasing energy efficiency with better insulating materials, and improving home security with locks and alarm systems. 240 pages, 282 illustrations. **Book No. 3248, $14.95 paperback, $22.95 hardcover**

DECKS AND PATIOS: Designing and Building Outdoor Living Spaces—*Edward A. Baldwin*

This handsome book will show you step by step how to take advantage of outdoor space. It's a comprehensive guide to designing and building decks and patios that fit the style of your home and the space available. You'll find coverage of a variety of decks, patios, walkways, and stairs. Baldwin helps you design your outdoor project, and then shows you how to accomplish every step from site preparation through finishing and preserving your work to ensure many years of enjoyment. 152 pages, 180 illustrations. **Book No. 3326, $16.95 paperback, $26.95 hardcover**

MAKING ANTIQUE FURNITURE
—Edited by Vic Taylor

A collection of some of the finest furniture ever made is found within the pages of this project book designed for the intermediate- to advanced-level craftsman. Reproducing European period furniture pieces such as a Windsor chair, a Jacobean box stool, a Regency table, a Sheraton writing desk, a Lyre-end occasional table, and many traditional furnishings is sure to provide you with pleasure and satisfaction. Forty projects include materials lists and step-by-step instructions. 160 pages, fully illustrated. **Book No. 3056, $15.95 paperback only**

FRAMES AND FRAMING: The Ultimate Illustrated How-to-Do-It Guide—Gerald F. Laird and Louise Meière Dunn, CPF

This illustrated step-by-step guide gives complete instructions and helpful illustrations on how to cut mats, choose materials, and achieve attractively framed art. Filled with photographs and eight pages of full color, this book shows why a frame's purpose is to enhance, support, and protect the artwork, and never call attention to itself. You can learn how to make a beautiful frame that complements artwork. 208 pages, 264 illustrations. **Book No. 2909, $12.95 paperback only**

Prices Subject to Change Without Notice.

Look for These and Other TAB Books at Your Local Bookstore

To Order Call Toll Free 1-800-822-8158
(in PA, AK, and Canada call 717-794-2191)

or write to TAB BOOKS, Blue Ridge Summit, PA 17294-0840.

Title	Product No.	Quantity	Price

☐ Check or money order made payable to TAB BOOKS

Charge my ☐ VISA ☐ MasterCard ☐ American Express

Acct. No. _____ Exp. _____

Signature: _____

Name: _____

Address: _____

City: _____

State: _____ Zip: _____

Subtotal	$ _____
Postage and Handling ($3.00 in U.S., $5.00 outside U.S.)	$ _____
Add applicable state and local sales tax	$ _____
TOTAL	$ _____

TAB BOOKS catalog free with purchase; otherwise send $1.00 in check or money order and receive $1.00 credit on your next purchase.

Orders outside U.S. must pay with international money order in U.S. dollars.

TAB Guarantee: If for any reason you are not satisfied with the book(s) you order, simply return it (them) within 15 days and receive a full refund.

BC